HOT ROD
volume 12
TECHNICAL LIBRARY

The Best of
HOT ROD
M A G A Z I N E

High Performance
CHRYSLER
Engines

CarTech
Auto Books & Manuals

Published by CarTech, Inc.
11605 Kost Dam Road
North Branch, MN 55056 U.S.A.
United States of America
Tel: 800-551-4754, Fax 651-583-2023

ISBN 1-884089-51-8

Book Trade Distribution by
Voyageur Press, Inc.
123 North Second Street
P.O. Box 338
Stillwater, MN 55082 U.S.A.
Tel: 651-430-2210, Fax 651-430-2211

Distributed in England by
Brooklands Books Ltd
P.O. Box 146, Cobham
Surrey KT11 1LG
England
Tel: 01932 865051, Fax 01932 868803

Distributed in Australia by
Brooklands Books Ltd
1/81 Darley St, P.O. Box 199
Mona Vale, NSW 2103
Australia
Tel: 02 9997 8428, Fax 02 9979 5799

Printed in Hong Kong

CONTENTS

Mopar Performance 528 Crate Hemi

The Hemi Is Back— But 102 More Inches Make It a Torque Monster!

By Marlan Davis

It wasn't enough that Mopar Performance (MP), Chrysler's performance-parts program, saw fit to bring back all the legendary 426 Hemi engine parts for the resto market. Now they've gone and developed a 528ci stroker version, the largest and most powerful factory crate engine you can buy. Since it's delivered completely assembled out of the crate, about the only parts you need to add are a carburetor, an ignition coil, plug wires, exhaust manifolds or headers, and accessory drive pulleys.

The foundation of this new-breed Elephant is a just-released 4.5-inch–bore siamesed cylinder block combined with a 4.15-inch–stroke forged crank. Despite the 0.400-inch–longer stroke (when compared to an original 426 Hemi), the engine remains internally balanced, utilizing a standard Hemi auto-trans flexplate and

The siamesed-bore block features cross-bolted main caps, a stock 10.72-inch deck height, and standard bearings. It accepts a stock oil pan, front cover, and Hemi heads. The bottoms of the cylinder bores are notched to clear the stroker crank (*inset*).

Photos by Marlan Davis

Able to shake the very foundation of society as we know it, the 528 Hemi makes 656 lb-ft of torque and 611 hp. The engine comes with an aluminum water pump and a damper; the pulleys are extra.

a degreed, SFI-approved, "thick" '66-'71-style damper. Forged-steel connecting rods hook the crank to forged-aluminum 10.5:1 pistons. New large-bore composition head gaskets seal the aluminum Hemi heads to the block. The heads' stainless steel 2.25-inch intake and 1.94-inch exhaust valves are actuated by a factory-engineered flat-tappet hydraulic cam working through the original-style Hemi shaft-mounted rocker arms. Designed for mild competition applications, the cam is compatible with both automatic or manual transmissions and spins via a roller timing chain. The exhaust valve heads are radiused for improved flow.

Topping off the heads is an M1® dual-plane, single four-barrel Holley-flange intake manifold. Never previously produced, it matches the power output of the old production dual four-barrel package yet improves driveability, throttle response, and low-speed torque. The engine's MP electronic distributor and "Orange-Box" electronic control unit offer a vast improvement over the original '60s points-type ignition system. Lubrication is provided by a 7-quart (with filter) '70-'71 E-body (Barracuda/Challenger) oil pan, windage tray, and high-volume oil pump.

The initial prototype engine was built and tested by the ace Hemimen at Dick Landy Industries. Essentially the same as the definitive production crate engines, this version differed only in the substitution of Manley capscrew rods in place of the still-under-development MP rods. To the basic crate package, Landy added 40 percent overdriven pulleys (actually a performance handicap, but great for guaranteeing cooling while cruising on hot summer afternoons). He also added dyno headers with 2¼-inch primaries and 4-inch collectors and a new Holley Pro Series HP 950-cfm double-pumper carb (PN 0-80496). Using 92-octane pump gas, 33 degrees total advance, and No. 78 primary jets in the carb, power peaked at an awesome 611 hp at 5,900 rpm, with the engine developing more than 500 hp from 4,100 through 6,200 rpm, the highest rpm point tested. Torque production was broader than the Pacific Ocean, with the mighty Elephant generating more than *600 lb-ft* (!) from 3,000-5,100 rpm and over 500 lb-ft be-

4

tween 5,200 and 6,100. The Hemi's torque peak of 656 lb-ft at 3,800 rpm was by far the most torque we've ever seen a crate engine generate. These numbers were also slightly better than MP's "official" 610hp, 555–lb-ft rating. Run—don't walk—down to your nearest MP dealer and grab hold of this menace to society. It should be available in late 1997 or early 1998. **CC**

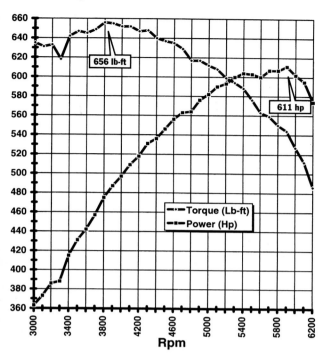

Mopar Performance 528 Hemi

656 lb-ft

611 hp

Torque (Lb-ft)
Power (Hp)

Rpm

Forged from 1053 steel, the 4.15-inch crank (PN P5249208) has hardened and radiused bearing journals. Note the offset of the stroker crank (*inset, right*) compared to a stock 3.75-inch–stroke Hemi crank (*left*).

Still under development as this is written, the crate engine's con rods are said to look similar to this factory rod assembly but will be made from stronger 8640 steel. Like this one, the new version is bushed for floating pins and has polished beams.

The stroker Hemi pistons use 1.031-inch–diameter floating pins and bulletproof round-wire retainers. Premium Speed-Pro ⅟₁₆-inch plasma-moly top, ⅟₁₆-inch ductile-iron second, and ³⁄₁₆-inch standard-tension oil rings are used.

Interchangeable with old cast-iron Hemi heads, the engine's aluminum castings (P4529336) are more than 50 pounds per set lighter. The combustion chambers are approximately 170 cc, the same as the original heads.

The valvetrain looks the same as it did in 1970, but new valvesprings can now handle up to 0.540-inch–lift cams. Improved valve-stem seals pirated from today's production Magnum V-8 keep oil out of the chambers.

MP 528 Crate Hemi (PN N/A)
Does not include plug wires, carb, starter, air cleaner, or exhaust manifolds.

Displacement:	528.0 ci (8.7 L)
Bore x Stroke:	4.50" x 4.150"
Power (as tested):	611 hp @ 5,900 rpm
Torque (as tested):	656 lb-ft @ 3,800 rpm
Block:	Cast-iron with cross-bolted four-bolt main bearing caps (positions 2-3-4), siamesed cylinder walls, blind head-bolt holes
Crankshaft:	Forged 1053 steel, eight-bolt flywheel flange, fully radiused main bearing journals
Connecting rods:	Forged 8640 steel, ⅜" rod bolts, 6.86" center to center, bushed pin end
Pistons:	Forged aluminum, 10.5:1 CR, floating pin
Camshaft:	Hydraulic flat-tappet (PN P4349257), 0.524"/0.540" lift (1.57:1/1.62:1 rocker arms), 292°/292° adv. duration, 108° lobe separation, 105° intake centerline
Cylinder heads:	Cast aluminum, adjustable valvetrain, 170cc Hemi chambers, 1 spark plug/cylinder, 2.25" intake/1.94" exhaust valves
Induction:	Holley 950-cfm double-pumper carb*, M1 dual-plane high-rise aluminum intake
Ignition:	MP race coil (P3690560)*, "Orange-Box" ECU (P4120505), MP vacuum-advance electronic distributor (P3690432)
Exhaust:	2¼" primaries x 4" collectors*
Warranty:	90 days parts from purchase date
Retail price (6/97):	N/A

*As tested; not included with engine assembly.

Sources
(*Product source*)

Mopar Performance Headquarters
Dept. CC
P.O. Box 360445
Strongsville, OH 44136-9919
Catalogs, tech manuals: 800/348-4696
Tech line: 810/853-7290

(*Test facility*)

Dick Landy Industries
Dept. CC
19743 Bahama St.
Northridge, CA 91324
818/341-4143

Building A 6,000hp Nitro Burner

By Gray Baskerville

The other day one of our readers took the time to drop us a note. He had seen Kenny Bernstein's Budweiser King Top Fuel dragster light up the clocks with a 4.60, 318-mph pass. His question was simple: What does it take to build an engine capable of accelerating a 1,900-lb vehicle from a dead stop to nearly 320 mph in just 1,320 feet? I didn't know the answer, but I know a guy who does—namely "Double-A" Dale Armstrong, the Bud King's crew chief and one of drag racing's most thoughtful innovators.

Top Fuel racing is unique because there are few restrictions to inhibit the engine builder. A cursory look at the rule book indicates that only the displacement, induction system (fuel injec-

tor and supercharger), cylinder head configuration, and electronics come under the National Hot Rod Association's no-no edict list (some entries, I might add, were the direct result of Armstrong's legendary imagination).

Unlike every other racing engine, long-term durability isn't a biggie—brute horsepower is! Like 6,000 *pissed-off Percherons* at 7,500, or like 5,000 *Sasquatch pounds of torque* at 6,000 rpm. "We bleed off far more launch-power than a Pro Stocker makes," says Armstrong, "and we do it just to keep the 17-inch-wide slicks from losing traction. We burn 58.86 cc's of fuel 60 times a second at 7,200 rpm. And I don't want to hear about a broken rocker arm slowing down a car. If we break a rocker the entire engine—block, heads, crank, rods, pistons, bearings, and blower—are, at best, greatly overstressed, and sometimes rendered totally beyond repair…they're history!"

History is how this whole "nitro" thing started. Back in the '30s, the Germans discovered that when they tickled a Grand Prix race engine's fuel mixture with a cleaning solvent called nitromethane, it produced a profound in-

crease in performance. A returning GI brought the nitro secret with him, and by the early '50s, nitro had become the "poor man's supercharger."

From a laboratory standpoint, the perfect air/fuel ratio when employing "pop" is 2 pounds of air to 1 pound of 100 percent nitromethane. So the evolution of Top Fuel has been one of forcing more and more fuel, and more and more air into an engine. The secret is how to burn most of it. "We're probably 1-to-1," concedes Armstrong, "but we want to use the unburnt 1 pound of fuel as a coolant. Horsepower comes from high cylinder pressure. However, the tremendous internal heat and cylinder pressure produced by nitro would melt the engine. For instance, a blown alky motor will produce more peak pressure than ours—but it won't continue that peak pressure the entire length of the power stroke like ours. This is because nitro burns slower than alky or gas but has a far greater expansion ratio. In fact, it continues to burn even at bottom dead center (BDC), and there is still a lot of residual cylinder pressure and unburned gas remaining when we open the ex-

What's Behind Blown Fuel

Power doesn't happen—guys like Bud King crew chief Dale Armstrong (right) and engine man Geoff Scarp *make* it happen.

The main saddles are align-bored because the clearances are critical. Better blowers have caused main bearing problems, so Armstrong now uses ⅝-inch main studs (left) and repositioned oil feed holes placed deeper in the main bearing saddles.

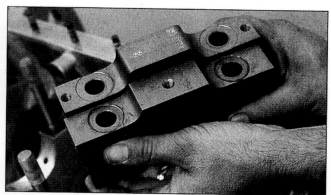

Another development is stronger billet aluminum main caps (background) with more meat in the center. An older 7076 cap is in the foreground.

Keith Black Racing Engines supplies the Stage-10 cylinder blocks for Armstrong's 6,000hp nitro motors. Cast by Alcoa and machined by Black, these aluminum blocks have been heat-treated for better yield, tensile strength, and elongation.

Sonny Bryant supplies the internally balanced, 68-lb billet cranks. The 4½-inch strokers last between 6 and 8 runs, then are pitched.

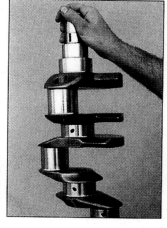

haust valves. This prolonged expansion acts to scavenge the unburnt exhaust gasses to such an extent that the burn is maintained in the exhaust pipes. This is why nitro motors are so loud."

Burning massive amounts of liquid dynamite forced into a blown Fueler has been just as revolutionary as using nitro. "We leave the line from a near-dead idle (2,500 rpm) burning 50 gallons of fuel per minute," says Armstrong. "Within .0250 seconds the rpm is up to 8,000 before the clutch begins a series of progressive lockups. At mid-track the rpm is down (7,000), the combustion chamber temps are up, and volumetric efficiency is increased to the extent that 62 gallons per minute can be burned. As the clutch fully locks up and the engine begins to climb to over 8,000 rpm, the volumetric efficiency begins to decrease. Fuel consumption is then 'pulled back' to 56 gallons per minute to keep from dropping cylinders."

With this in mind, the real stepping stone to 300 mph and beyond has been making the spark stronger. Two fuel pumps, two barrel valves, three sets of nozzles, huge ports in twin plug heads, big blowers, and lots of boost don't mean diddily squat if the magnetos can't ignite the fuel mixture. For the past five years, Armstrong has been working on magnetos to produce more of what they call "joules of energy" at the plugs them-

selves. Armstrong was one of the first to successfully employ two magnetos—he even messed with three but that innovation (along with variable speed blower drives, and electronically actuated clutch lockup devices) became NHRA no-nos. First, he messed with rare earth magnets that were more powerful than the ones contained in the then-current magnetos. He also increased the capacity and altered the location of the condensers so they wouldn't burn up the points. The net effect was the first 300-mph run.

Because a Top Fuel engine can be likened to a grenade with a 4.5-second fuse, its internal components and lubricants must be stronger than a billy goat's

AT A GLANCE

Engine: 90-degree V-8 with 4.189-inch bore, 4.500-inch stroke, 497ci displacement; 40-inch height, 33-inch width, 28¾-inch length; estimated 6,000hp at 7,500 rpm; estimated 5,000 lbs-ft torque at 6,000 rpm; $45,000-50,000 building cost
Best e.t.: 4.59 seconds
Best top speed: 318.69

Cylinder Block: 8-cylinder Keith Black Stage-10 wet style; 4.800-inch bore spacing; 146-lb weight; tapered cast ductile iron liners (.175-inch top, .165-inch bottom) honed with 600-grit stone

Crankshaft: Sonny Bryant 90-degree billet steel; internally balanced with no center counterweights; 68-lb weight; .095-inch bearings (tin over babbitt featuring soft material for improved "embedability")

Connecting Rods: BRC, 7075 T6 forged aluminum; various lengths in .020-inch increments (6.981-inch shown here); 890-gram weight; .065-inch tin-over-babbitt rod bearings

Pistons: Venolia forged aluminum; staggered compression ratios from 5.75:1 to 7.5:1 flat-top with intake valve relief; 760-gram weight; coated with heat barrier and lubricant to inhibit skirt galling from raw fuel; Dykes top rings; Moly middle; double-rail expander bottom; .028 to .024-inch end gap

Piston Pin: Dynamic Machine 1.555-inch full-float; 332-gram weight; 3.300-inch length tapered to ½-inch i.d.

Lubrication System: System One wet sump gear type; 22-24 gallons-per-minute capacity; 200 psi maximum oil pressure at idle, 195 psi in lights; 11-quart Moroso aluminum pan

Camshaft: Erson dual-pattern, 9310 steel roller with .720-inch intake, .660-inch exhaust lift at valve; 302 degrees intake, 296 degrees exhaust with .050-inch of lifter rise; 112-degree lobe centers; 90-degree overlap; 2-degree advance; Casale gear drive

Valvetrain: Roller tappets; Jesel pushrods; 10.650-inch intake length, 11.400-inch exhaust; ⅜-inch diameter intake, ½-inch exhaust; Billet chrome-moly rocker arms; 1.54 intake ratio; 1.57 exhaust ratio; .024-inch lash; Erson titanium retainers; Dechellis Machine rocker arm stands

One of the unusual aspects of Top Fuel racing is that engines are tuned by juggling the compression ratios from front to rear. This is done by the heights of the pistons and the length of the connecting rods. Pistons are coated with a thermal barrier/lubricating material.

Armstrong believes that you can't have too much oil pressure, so he relies on a System One gear-type oil pump with a pumping capacity of 22-24 gallons per minute. Eliminating the oil filter saves 5 pounds.

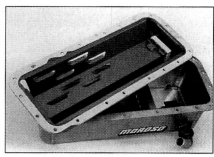

The Moroso-fabricated aluminum pan contains 11 quarts of 70-weight Prolong oil. A dry-sump system is not used because Armstrong claims they are too heavy and hard to clean between rounds should a failure occur.

The Erson dual-pattern roller tappet cam has run with 2 degrees advanced. The 9310 steel shaft is strong enough to survive a direct blow from a wayward piston or rod without breaking. Casale supplies the gear drive.

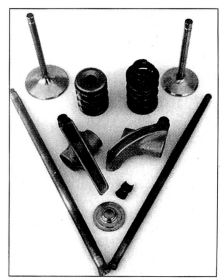

Blown-fuel engines produce long-term cylinder pressure, so opening the exhaust valve even at BDC is monumental. Consequently all the valves and valve springs are checked after each run, and the exhaust valves exchanged for new ones.

breath. Blocks with huge main webs, chrome-moly studs, billet aluminum main caps and steel cranks, forged aluminum rods and pistons, cams and valvetrain components begin to tell the story. These parts must have the capacity to withstand 10,000-rpm runaway speeds when driveline components fail, including oil pumps generating 200 psi of oil pressure at idle and 195 psi in the lights. These engines also use a new oil specifically formulated for Top Fuel racing that's all part of producing the 6,000hp act. As for the oil developed by Prolong

Intake Valves: Manley titanium; 2.375-inch head size; ¹¹/₃₂-inch stem diameter; 1.29-gram weight

Exhaust Valves: Manley Inconnel; 1.975-inch head sizes, ⅜-inch stem diameter, 1.66-gram weight

Intake Valve Springs: Speed-Pro double with dampener type; 260-270-lb closed seat pressure

Exhaust Valve Springs: Crane triple-wound, 320-340-lb seat pressure

Cylinder Heads: Alan Johnson billet aluminum; 170cc combustion chamber; 2.285 by 2.200 intake ports; 1.500 by 2.250 exhaust ports; 13¾₆-inch in head, four in valley head studs; 120-130lb torque; Clark copper .093-inch head gaskets

Intake Manifold: Keith Black magnesium; 22-lb weight; ported to match cylinder heads

Supercharger: SSI Roots type; 14-71 with 19-inch Teflon-tipped rotors; 40 psi (in the lights) maximum boost; 25.8. percent overdrive; 3-inch drive belt

Fuel System: Waterman two-segment; 62 gallons per minute; 500-psi fuel pressure; 10 nozzles in injector, 8 in intake runners, 16 in cylinder heads

Fuel Injector: 12½-lb Carbon Speed; three 5-inch butterflies inlet size with a total of 62 square inches maximum; Pete Jackson barrel valve

Ignition System: 44-amp, 800-volt MSD with 50 degrees advance lead; MSD 8mm suppressor core plug wires; NGK No. 10 plugs; .045-inch gap

Headers: 2½-inch tube diameter; mild steel; 18-inch length

to cushion the Bud King's crank, Armstrong swears by the stuff: "I can't tell you how good it is and how it saved our ass last year."

What's in store for this year is three-fold. Refinements in his engine's internal dynamics (i.e., different rod lengths and strokes), clutch components, and control of tire shake lead the list. Even though Kenny Bernstein's Bud King *is* king, you can bet your sweet bippy that Double-A Dale isn't going to stand around and wait for the next footfall—he's gonna make it fall. Do we hear, see, and feel 320? How 'bout 4.4? Another Number One? More NHRA no-nos? We'll keep you posted.

Billet aluminum cylinder heads are far stronger than those with water jackets. Each of these 32-lb beauties, machined by Alan Johnson, has been fitted with 16 down nozzles positioned just behind the intake valve.

The head features 170cc combustion chambers and two spark plugs to help burn the incredible amount of fuel that is force-fed into them. It has been O-ringed, as well.

The intake and exhaust ports reflect Armstrong's "more-and-more" philosophy. The mouth of the intake port is 2.284 inches wide and 2.200 inches tall, while the exhaust ports measure 1.50 inches tall and 2.250 inches wide.

The induction system is comprised of a 22-lb Keith Black magnesium intake manifold, an 82-lb, SSI-prepared 14-71, Roots-type supercharger, and a 12½-pound Carbon Speed carbon fiber injector fitted with a Pete Jackson barrel valve. A 3-inch drive spins the blower at 25.8 percent over.

Armstrong was the first to employ two pumps and two barrel valves. A two-segment Waterman fuel pump—rated at 62 gallons per minute—maintains the fuel pressure at 500 psi max. Fuel delivery is 20 percent through the 10 injector nozzles, and 80 percent through a combination of 8 port nozzles and 16 down nozzles.

Lighting the load is left to a pair of MSD magnetos capable of generating 44 amps and 800 volts per unit. Each is locked out at 50 degrees, and sends the spark via MSD 8mm suppressor-core wires to Number 10 NGK plugs. **HR**

DREAM
MOTORS

KILLER
ENGINES
FOR MONDO
POWER

STREET HEMIS

By David Freiburger

The Mopar Hemi is a cliché of horsepower. The early FirePower-era hemroids were the beginning of a speed legacy that later evolved into the mythical 426. When the new Hemi took a First/Second/Third sweep at its 1964 Daytona 500 introduction, it was an indication of the performance that later caused NASCAR to outlaw the engine. That move sent Richard Petty to the dragstrip, where Hemis also thrived.

Mopar Elephants were the first to go 200, 250, and 300 mph at the digs, and every Top Fuel championship has been won by a Mopar-derived Hemi except in 1966, when Pete Robinson cleaned up with his SOHC Ford. Factory-built Hemi door-slammers still own the crowd-pleasing Super Stock/A classes, which are now producing e.t.'s in the 8s. Hemis also breed on the salt flats, where a pachyderm Chrysler took the record for the world's fastest piston-driven vehicle when Al Teague's Speed-O-Motive streamliner ran 409.846 mph in August of 1991.

Since early hot rods were boulevard clones of race cars, the Hemi quickly became a stoplight legend, too. And now that street horsepower is making a comeback, we're glad to say that killer Hemis are back in vogue; they're found in street rods and street machines, twisting four-speeds and punishing tires. Read on for some of the latest Hemi news and some parts sources so you can start building the Hemi of your dreams.

WHY ARE HEMIS BETTER?

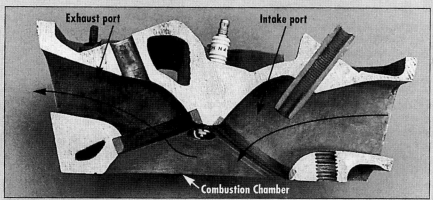

Exhaust port   Intake port

Combustion Chamber

The term "Hemi" is an abbreviation of "hemispherical combustion chamber," which is the design feature that makes the engines relatively unique. While there is no firm engineering definition of a Hemi, it is generally accepted to be a half-sphere-shaped combustion chamber with the valves directly opposite and inclined away from each other and the spark plug centered in the chamber. In 1966, W.L. Weertman, Chrysler's Assistant Chief Engineer of Engine Design at the time of Hemi development, summed up the benefits of the design: "(1)

The air can flow straight from the intake port into the chamber and straight out the exhaust. (2) The intake valve tips in the direction of its port and this makes a uniform air distribution at the valve head. (3) There is no shrouding of either the exhaust or intake valves when the valves are open. The only exception in this design is the proximity of the bore wall to the edge of the intake valve. (4) The spherical shape gives us the well-known plus on thermal efficiency. (5) The spark plug is almost right on bore center for an additional plus on igniting the charge."

BUILD A BRAND-NEW ELEPHANT

A good indication of the Hemi rage is the fact that Chrysler Corporation has supported its heritage by releasing the Mopar Performance (MP) line of 426 Hemi parts. Hemi blocks, cranks, manifolds, and all the small parts are available direct from your local MP dealer. By the time you read this, both iron and aluminum reproduction Hemi heads will also be available. That means that you will be able to put together a complete Hemi using entirely brand-new Mopar parts. MP even offers a Hemi block with wedge motor mounts to make it much easier to install. For more information, contact Mopar Performance, Dept. HR08, P.O. Box 215020, Auburn Hills, MI 48321, 313/853-7290 (tech line).

KEITH BLACK STREET HEMIS

One of the biggest names in Top Fuel Hemi engines is Keith Black (KB), and while Keith passed away two years ago, Keith Black Racing Engines is still going strong. In fact, it's attacked the street market with a line of street Hemi blocks and heads. The blocks are similar to the Top Fuel units but use dry sleeves and water jackets, and they're plumbed for hydraulic cams. The KB Hemi also uses ductile-iron main caps and is optional with six-bolt mains. It comes with provisions for a stock oil pump, but you have to use an external oil pickup. You also need to specify if you want a fuel-pump boss and a dipstick hole, but everything else can use stock Hemi parts, including the motor mounts. However, if you want to use stock iron heads, you'll have to plug and redrill the steam holes in the heads. The bore size is optional, but the block can handle a 4.500-inch bore and a 4.500-inch stroke for 572 cubic inches. Contact Keith Black Racing Engines, Dept. HR08, 11120 Scott Ave., South Gate, CA 90280, 310/869-1518.

The KB wet heads can be used on a stock iron block or the KB aluminum block. Black also offers complete valvetrain parts, valve covers, and gear-drive units.

Keith Black can supply a streetworthy aluminum block and a trick steel crankshaft, but the company recommends Carillo steel rods for the street. Keith Black Signature Series hypereutectic pistons for street Hemi will also be available soon. The company has all the tech support and parts you'll need to put together a real stomper Hemi.

EARLY HEMIS ARE BACK

There's a certain crowd that'll take the FirePower, FireDome, and Red Ram early Hemis over the late-model 426 stuff any day. They are the guys who appreciate vintage speed with classic styling. Of course, a 6-71 blower is required, but few motors can eclipse the appeal of an early Mopar Hemi in a street rod. And they can really scoot, too!

The most popular early Hemis are the Chrysler FirePower 331, 354, and 392. While the early Hemi can be cheaper to buy than a 426, it is almost an entirely different engine, so the parts are harder to find. However, we did some networking and came up with a great list of suppliers for such goodies as electronic distributors, blower parts, rebuild parts, accessory brackets, and tranny adapters for TorqueFlites, TurboHydros, and four-speeds. We even discovered that Donovan has

rereleased the 417 early-Hemi-based aluminum race block from the early '70s! Another hot tip is that 440 connecting rods and rod bearings will fit in a 392 if you widen the rod journals (or narrow the rods) and run custom pistons from a company like C&A, JE, or Venolia. In fact, all the early Hemis had steel cranks, so you might as well stroke that sucker while you're at it! These tips and more can be had from some of the early-Hemi-specific suppliers listed.

Chris Karamesines used a fuel 392 to run the first 200-mph quarter-mile in 1960, although some say he was mixing hydrazine. It was 1964 before Don Garlits would repeat the performance officially at Island Dragway in New Jersey, also with a Hemi.

An early Hemi just doesn't look right without a blower. Both Weiand and BDS have all you'll need to huff your FirePower Hemi.

Quality Engineered Components makes adapters *(arrow)* so you can use an A-motor TorqueFlite or four-speed on an early Hemi, and Hyatt Engineering has adapters for GM trannies. Also note the Quality Engineered Components electronic distributor.

Weiand is celebrating its 60th anniversary this year, so early Hemis are nothing new to the company. Weiand still has dual-quad intakes, blower manifolds, timing covers, and Chevy water-pump adapters.

EARLY HEMI PARTS SOURCES

Blower Driver Service
Dept. HR08, 12140 E. Washington Blvd.
Whittier, CA 90606, 310/693-4302
Blowers and blower drives

Donovan Engineering
Dept. HR08, 2305 Border Ave.
Torrance, CA 90501, 310/320-3772
417 aluminum race blocks, race accessories

Egge Machine
Dept. HR08, 11707 Slauson Ave.
Santa Fe Springs, CA 90670,
310/945-3419
Rebuild kits, pistons, bearings

Hyatt Engineering
Dept. HR08, 547 Sinclair Frontage Rd.
Milpitas, CA 95035, 408/946-2007
Accessory brackets, trans adapters, Chevy water-pump adapters, rocker stands

Mallory, Inc.
Dept. HR08, 550 Mallory Way
Carson City, NV 89701, 702/882-6600
Plug wires

Quality Engineered Components
Dept. HR08, 1150 Ryan Ct.
West Linn, OR 97068-4034,
503/656-4545
Early Hemi ID info, trans adapters, internal components, electronic distributors

Reath Automotive
Dept. HR08, 3299 Cherry Ave.
Long Beach, CA 90807, 310/426-6901
High-performance, full-race, and rebuild parts

Sanderson Headers
Dept. HR08, 202 Ryan Way
South San Francisco, CA 94080
415/583-6617
231, 330, 331, 354, and 392 block-hugger headers

Weiand Automotive Industries
Dept. HR08, 2316 San Fernando Rd.
Los Angeles, CA 90065, 213/225-4130
Intakes, blowers, and drives, water-pump adapters

EARLY HEMI IDENTIFICATION

Thanks to Gary Stauffer at Quality Engineered Components for supplying this information on how to recognize some of the early Hemis. Remember that you could find a Chrysler Hemi in a Dodge truck or even in a combine or an agricultural water pump!

MAKE	CID	CODE
Dodge	241	D44, D50
Dodge	270	D55
DeSoto	276	S16, S17, S19
DeSoto	291	S21, S22
Chrysler	301	WE55
Dodge	315	D500
Dodge	325	KD500
DeSoto	330	S23, S24
Chrysler	331	C51, C52, C53, C54, NE55, CE55, 3NE55, WE56
DeSoto	341	S25, S26
Chrysler	354	NE56, CE56, 3NE56, WE57, LE57, 58W, 58S
Chrysler	392	NE57, CE57, 3NE57, 58N, 58C, 58N3

HEMI HEADS FOR 440 BLOCKS

If you've just gotta have a Hemi, one of the most affordable ways to do it is with Stage V Hemi conversion heads. While stock Hemis have a different head-bolt pattern and water-jacket location than the wedge blocks, Eric Hansen at Stage V has solved those problems with an aluminum head that is designed to bolt onto 361/383/413/426/440 wedge motors. The 440-type block is a lot easier and cheaper to come by, and is also easier to mate to a K-member. The Stage V heads have better flow than stock Hemi units and use 2.200 or optional 2.250 intake valves, 1.900 exhaust valves, and a special rocker-arm setup with 1.6/1.4 ratios that Eric says improve valvetrain geometry and allow you to use off-the-shelf cams with his heads. Remember that Hemi conversion heads also require Hemi pistons, intake, and headers or manifolds. Stage V also makes regular Hemi heads for both street and race use. The conversion heads are water-cooled versions of the Stage V Top Fuel and Alcohol heads that Frank Bradley used to run the first 4-second e.t. on the West Coast, and that Al Teague uses on his streamliner. For info, contact Stage V Engineering, Dept. HR08, P.O. Box 827, Walnut, CA 91788-0827, 909/594-8383.

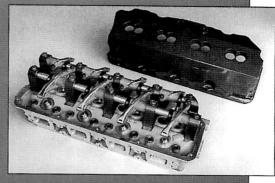

Stage V makes Hemi heads that work with 440 blocks. They aren't available with dual plugs because of the custom valvetrain required, but the magnesium valve covers are cool.

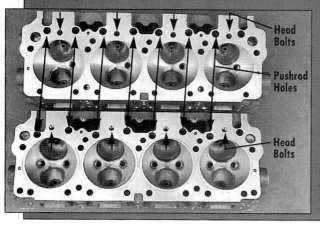

Head Bolts

Pushrod Holes

Head Bolts

The underside is where you see the main difference in the standard Hemi head (bottom) and the conversion head (top). Note that the pushrod holes and indicated head-bolt holes are altered. The conversion is also available with a better exhaust-flange bolt pattern for better sealing, and Stage V sells headers for the raised-exhaust-port version of the heads.

With the Stage V heads and the plethora of repop parts available, you can build a Hemi race car from scratch. Mosher's Muscle Car Motors (Dept. HR08, 11930 Sheldon St., Sun Valley, CA 91352, 818/504-9010) sells simulated Hemi race cars like this 11-second, 500-inch '65 Belvedere. It's a driver!

INSTALLING A HEMI

Hemi motor mounts and frame mounts are not common with any other engine, but there are several ways to install a 426 Hemi in your Mopar musclecar. For A bodies you'll need a set of Elephant Ears, which are aluminum

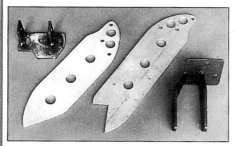

From left to right are a Mickey Mouse mount, a set of Elephant Ears, and a Battleship mount, which is a solid left-side motor mount for a Hemi block on a Hemi K-member. They are all available from Chrysler Performance Parts Association, Dept. HR08, P.O. Box 1210, Azusa, CA 91702, 818/303-6220.

brackets that bolt behind the water pump and extend to the frame (see photo). For a 426 Hemi swap into a B or E body, your options are:

(1) Use an original Hemi K-member, mounts, and brackets with an original block or a Mopar Performance part No. P4529850 Hemi block.

(2) Use a Mopar Performance block part No. P4529852. It's a Hemi block with wedge-style motor mounts, so it will bolt into a standard small-block or big-block B or E body.

(3) Use a standard 440 block with Stage V Hemi conversion heads (see "Hemi Heads For 440 Blocks").

(4) Use a Chrysler Performance Parts Association Mickey Mouse mount. That bracket allows you to bolt a Hemi block to a standard V8 K-member on the left side. The right side of a Hemi block has ears cast into it that can be drilled to accept a standard 383/440 motor-mount bracket.

426 HEMI PARTS SOURCES

The demand for 426 Hemi power means that there isn't as much of a parts shortage as there was a few years ago. Many companies make Hemi speed parts, and Mopar restoration companies have items such as brackets, valve covers, and manifolds. Here's a list of some 426 Hemi-specific suppliers that can either build an engine for you or help find the hard parts you need. **HR**

Dick Landy Industries
Dept. HR08, 19743 Bahama St.
Northridge, CA 91324, 818/341-4143

Hale Performance
Dept. HR08, P.O. Box 1518
Van Buren, AR 72956, 501/474-5252

McCandless Performance
Dept. HR08, P.O. Box 741
Graham, NC 27253, 919/578-3682

Ray Barton Engines
Dept. HR08, 7 Belle Alto Rd.
Wernersville, PA 19565, 215/670-8591

Cubic inches aren't everything, but they help. For years, they helped two-ton factory luxotourers roll down the road, and they· also helped make the supercar era extra exciting. And even though muscular mega-inchers are, for the most part, socially outcast today, in select applications and configurations they still have their place as premium performers.

Torquey V8s with generous displacement numbers are a pleasure to drive, and regardless of the weight they're asked to haul around, monster· motors always appear to make effortless power. Cubes offer full steam at low rpm, and they're just the ticket for a long-geared cruiser. Big-inchers just seldom seem to work very hard at all.

Of course it all their bottom-end twist were put to work in a more brutal fashion, a lightly stressed and durable bracket motor could result too. By relying on a broad torque output band rather than high engine speeds, big-inch powerplant internals are subjected to less wear and tear. Quite naturally, the rest of the drivetrain has to be up to the task of handling a big motor's gaff, but a mega-inch mill is theoretically capable of living a long and productive life without ever seeing a rev over six grand.

Almost all major automakers' assembly-line iron—the 500-cube Caddy, the 460 Ford, the fat 454 Chevy rat, the 440 MoPar and the three 455s from GM's B-O-P divisions—can supply the dependable repeater action required to get an ET runner back down the track time after time. But the simple old adage, "there's no substitute for cubic inches," can now be looked back upon as a passing phrase. Substitutes in the form of efficient mechanical combinations and flow-promoting hardware have evolved into more than adequate replace-

ments for inches. Thus far, however, there have been few efforts to mate the big-cube/high-flow approaches to going fast. So while stroker motors have been treated to reworked heads, and reworked heads have been screwed onto popular big-inchers, combining the best of both worlds has basically been ignored.

We recently came across one straightforward approach to a well-fed monster motor, and, we thought it would be interesting to take a closer look. The buildup was being handled by noted Michigan MoPar flogger John Tedder, for fitting to a Plymouth Barracuda used in both occasional street and strip action. The car has already been run with a reworked 400 wedge on board, but freshening-up plans for this season have been upgraded to include a punched-out and stroked version of the same basic engine. We're going to take a broad look at the overall powerplant assembly, with emphasis on the two main areas critical to efficient high-volume air handling. Since an internal combustion engine is little more

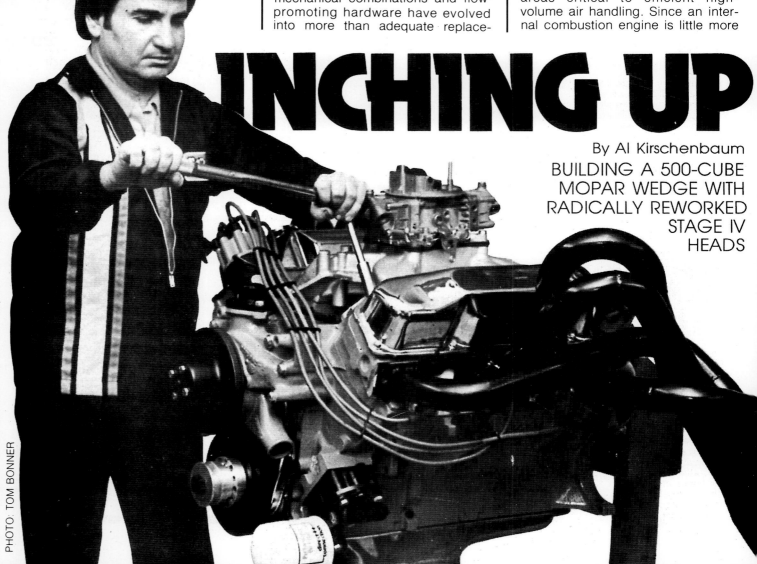

INCHING UP

By Al Kirschenbaum
BUILDING A 500-CUBE MOPAR WEDGE WITH RADICALLY REWORKED STAGE IV HEADS

INCHING UP

than an air pump, increasing the displacement without improving the point of maximum air flow restriction—the cylinder heads—must be considered a half-hearted effort. Tedder was responsible for building this 505-cubic-inch motor, while Corky's Competition (21047 Schoenherr, Warren, MI 48089) handled the Stage IV cylinder head prep, and Johnny's Custom Crankshafts (27257 Brest Road, Taylor, MI 48180) machined up the half-inch stroker crankshaft.

The foundation of the stroking process is, of course, a radically re-worked crankshaft. A stroker motor also involves specific design and machining considerations in the piston and engine block departments, as well as special attention to the final balancing of the reciprocating

A billet crank will run up a thousand-dollar tab at the shaft shop, and some engine builders still feel that forged steel production units are metallurgically stronger. Solid billets are the big-buck approach to big-inchers, and we're dealing with the basics, so these offset journals were formed with weld buildup and remachining. This lower-buck approach to stroking also eliminates the expensive centerweighting process.

Extending an arm requires additional room to move inside the engine block. The 440 casting was notched at the bottom of each cylinder bore (arrows) where the rods' big end hardware passes close by. Beefier aluminum rods or oversized extra-duty nuts and bolts will require even more clearance in this area. Since a number of other relationships are also altered by the longer stroke, clearance checks should be made between the counterweights, piston skirts and pin bosses, and the number three main bearing thrust flanges.

assembly; but shaft-crafting is the engineering axis around which most mega-inchers revolve.

Carving an ''arm'' is by no means a backyard project, and it's a task best left to professionals. In addition to the already complicated regrinding process, Chrysler crank journals (as well as those found on

Rebalancing the stroker crank required considerable quantities of heavy Mallory metal in the counterweights.

many other premium production shafts) were factory Tufftrided, a surface treatment that makes re-working even more difficult. (Tough outer bearing surfaces result from Tufftriding's .0002-inch-thick layer of nitrogen needle-impregnated steel which must be ground away before any welding can be done.) The increase in the crank's stroke length results from regrinding new rod journals on relocated, offset centers (in this case, .250-inch, or half the total stroke increase, away from the shaft centerline). There are specific instances where large-diameter rod journals can be re-ground smaller on new centers without a buildup of welded-on ma-

terial, but these applications are few and far between.

In order to compensate for the increase in iron hanging out at the edge of the crankpin, existing counterweights have to be drilled out and filled with extra-heavy Mallory metal, or they must be rebuilt and reshaped with wraparound welded-on material buildup. The most professionally proper plan for counterbalancing an extended arm is centerweighting—actually adding two new counterweights adjacent to the normally unweighted center main journal. All these methods help to reduce loading on the main bearings. Billet stroker cranks are machined from solid steel stock, so all these construction factors can be built-in to begin with.

Our 505-incher uses stock-weight connecting rods and trim-but-tough aftermarket pistons, so rebalancing was accomplished by adding heavy metal to the existing counterweights. Final finishing of the half-inch-longer arm involved cutting fresh radii at the rod journal cheeks, and remachined oil holes and chamfers, as well as numerous Magnaflux checks, along the way. The big wedge's crank was attended to at Johnny's, where their stroking tab was a modest $300. In contrast, a fully machined forged billet crank from the same shop (hewn from a single piece of solid steel) goes for around $1000.

Lengthening the crankshaft's stroke by a half-inch also extends the piston's travel by half that amount; and if an attempt were made to utilize the stock 440 slugs, they'd end up too far above the block deck at top dead center (TDC) and too far below the bore (and possibly into the path of the crankshaft) at bottom dead center (BDC). Since Tedder retained the standard-length connecting rods (6.76 inches), new pistons with re-located pins were required. John specified a set of forged Venolia

INCHING UP

lightweights with suitably lightened pins, located approximately .250-inch higher than stock, and accommodations for 1/16-inch compression rings. These slugs incorporate a pair of relatively shallow valve notches, and on assembly, the resultant deck height produced an even 9:1 compression ratio.

Sealed Power piston rings were specified in a .035-inch-over-size

Internal components include thoroughly massaged stock connecting rods, special-order Venolia pistons, lightweight pins, double Tru-Arc pin retainers and moly-filled Sealed Power rings.

configuration to allow custom end gapping. The plasma-sprayed moly-filled tops were set at .016-inch, while the channel-filled reverse-twist second rings were gapped at .010-inch. John feels that these tighter clearances are well suited to both the rings' lower placement on the pistons and the overall low compression ratio of the engine and the reduced thermal levels that the rings will see.

Stroking Chrysler's 440 wedge and retaining the original equipment (OE) connecting rods actually works out to the powerplant's advantage. Altering the ratio of connecting rod length to stroke length in this case produces more favorable numbers: 1.8:1 with the stock 3.75-inch stroke and 1.69:1 with the extended 4-inch arm. Reducing the L/R ratio in this manner is favorably suited to a number of aspects of this engine design, includ-

ing the increase in induction flow area, the lower rpm range and the wider torque band. In addition, the lower ratio is better suited to the automatic transmission application that the motor's intended for.

The OE connecting rods are the slightly stronger-than-standard "Six-Pack" beams fitted with high-strength Direct Connection ⅜-inch hardware (part No. P4120068). The rods are ground smooth and shot-peened; and in a departure from common race prep practice, the small ends remain unbushed, but they're opened up to "float fit" the piston pins. Aside from lower machine shop bills, the theory behind the unbushed design leaves a little more meat in the small ends that would normally be machined away for the bushings. A single oiling hole is drilled to lube the pin (fitted with .0015-inch clearance in the rod), and Tedder suggests using a very high-quality lubricant changed as often as possible.

Right out of the box, the Stage IV cylinder heads (part No. P4120352) flow approximately 20 percent better at .600-inch lift than a similar 1967 high-performance 440 production casting. Box stockers seem to fall off badly above .550-inch lift, and at that point, the new heads are only about 7 to 8 percent better. In finished form, however, comparable castings differ by as much as 15-17 percent.
Corky's Competition spends roughly 40 hours massaging a pair of Chrysler's Stage IV castings for maximum performance. Externally, signs of reworking are the visibly enlarged intake and exhaust ports seen in these side-by-side comparisons. Stock "Stage" heads are on the right.

Stage IV cylinder head preparation, even in view of this rather exotic application, must be considered conservative. The main aim here was for maximum intake port cross-section without breaking the head, either during preparation or later while in use. Corky's found that most of the air flow improvement in these castings comes through work in the under-valve bowl area. Chrysler improved the Stage IV pieces by "helping" the port mouth width at the short side radius. They also streamlined the back of the bowl, where there used to be a fairly large bump. This particular section has been straightened considerably. The short side radius was lowered and filled, and in line-of-sight terms, the entire port has been straightened out and the short wall is lower by about .370-inch.

Corky feels that air flow will dictate the ultimate horsepower of the motor, and the size of the port dictates the speed at which that power level is achieved. The larger the engine's displacement, or the higher the rpm it sees, the bigger the port area required. If the bowl area is

Most improvements in air flow through the Stage castings are found through work in the bowl areas beneath the valves (arrows). The last half-inch upstream of each valve is most critical.

made too large for a smaller wedge engine application, however (as in a 383 automatic, for example), the engine could "lay down" or run flat at low speeds. The bowl has more effect on air flow than any other place in the port.

The majority of flow improvements will be found in the last half-inch upstream of the valve and in the bowl. Most porting and/or polishing work here is mostly "sex appeal," but for a big enough displacement, or a high enough rpm, these areas should be enlarged considerably.

For this long-stroke, automatic-backed combination, the head's runner displacement was increased considerably (by about 9cc) over a standard big-block ported Stage IV casting. Corky mentioned that the cross-sectional areas could have

INCHING UP

been opened up even further, but in view of the engine's intended use, metal removal was approached conservatively.

The exhaust port mouth shouldn't be enlarged much at its exit point, and a look at the photos will show that aside from a .030-inch clean-up, no flaring in this area is attempted. These heads actually *prefer* a mismatch at the entrance to the exhaust header, and Corky cautions against any attempt at port or gasket matching on the outlet side of the heads.

As with all other Chrysler wedge heads, unshrouding the combustion chambers around the valves promotes flow. By making all chambers as wide as possible, and by cutting large, full radii in the corners (right up against the edges of the gasket), waste gas extraction is especially enhanced. In fact, this is the reasoning behind the use of the smaller 1.81-inch rather than the 1.88-inch valve size. This plan also applies to the smaller (4.25 as opposed to 4.32) 383-cube bore size. The standard MoPar scheme of notching the top of the cylinder block in these areas (around the valves) no longer applies, due to the higher ring placement on modern pistons. These portions of the bore can, however, be gently "feathered" for an unshrouding ef-

Camshaft selection for this oddball combination is wide-open, but there is a wide range of compatible Direct Connection parts to choose from. This 505 uses a 332-degree DC mushroom tappet grind (part No. P3690558), in conjunction with OE adjustable rockers.

fect, but a .060-inch minimum clearance must be maintained above the top rings.

Corky also points out that pushrod clearance is critical, with room being made at the external passage walls with a die grinder and a carbide burr. A thorough mock-up o'

Valve guides (arrows) are machined down ¼-inch for stem seal and retainer clearance, with the Perfect Circle seals taking up .155-inch of the OE guide meat and the remainder being allowed for guide-to-retainer clearance at maximum lift (.060-inch is the minimum suggested clearance at this point).

For that mild-mannered underhood look, a single 850-cfm Holley carb was modified for use on an extensively reworked "sleeper-style" intake manifold. The high-flow cylinder heads and the enlarged engine displacement dictate an intake runner volume increase on the order of 20-25 percent. For more serious bashing, Tedder keeps a 1300-cfm tunnel ram system in reserve.

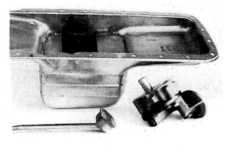

Careful component selection and meticulous assembly allows the use of a virtually off-the-shelf oiling system. Tedder fitted an 8-quart Moroso oil pan (No. 2076) and extended pickup (No. 2475) with a high-volume Direct Connection oil pump (part No. P4007177) and a hardened DC pump driveshaft (part No. P3571071). Frequent lube changes are especially important with this engine's steel-to-steel piston pin fit.

these related valvetrain pieces is suggested prior to machining and/or assembly. Most other wedge power systems follow proven competition plans as Tedder applies many ideas from his "Mr. MoTech" SS/AA hemi 'Cuda to this more

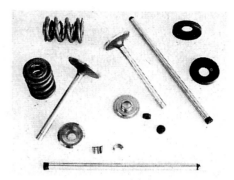

Exhaust valve size for big-inch B wedges has been found to be most advantageous at 1.81 inches, derived from cut-down versions of the OE 1.88-inch Max Wedge pieces. Since these earlier production valves have stems .0015-inch smaller than standard valves, bronze-wall guides are mandatory. Stem-to-guide clearances should be .0005-inch on the exhaust side and .0002 on the intakes.

The Manley spring seats seen at the upper right are specially hardened to prevent the valve spring and damper from cutting into the cylinder head. These pieces also keep metal chips out of the oil, and help stabilize the valve springs and extend their useful lives. Spring shims fit below the cups.

Pushrods are custom-built from a Direct Connection kit (part No. P4007284). Overall length is determined by centering the adjusting screw in the rocker arm and cutting the pushrod shaft to suit. Finished rods should have a minimum of 2½ to 3 threads showing below the rocker.

Most common keys (left) have been produced with what's known as a 7-degree plunge angle, and most valve spring retainers are machined to match this design. Current practice, however, has stepped up to 10-degree locks, and used with complementary retainers (right), the increased bearing surface area adds up to better load distribution and an almost nonexistent failure rate.

street-oriented piece.

The consequences of all these high-flow, mega-inch antics will become evident when Tedder and friends slip this monster motor into Merrill Bumstead's 3100-pound killer 'Cuda and assault the Milan, Michigan, quarter-mile. The added displacement alone should push the little Plymouth well beyond last season's mid-10-second marks, but coming up with the cubes was really the easy part. What remains is maintaining the mechanical pace, and preventing the others in the Motown supremacy race from catching on and inching up. **HR**

318 STREET RECIPES

NEW MOPAR POWER MEALS FEED HUNGRY 318s

By Marlan Davis

For years, Chrysler's Mopar Performance/Direct Connection folks have been refining their coordinated "LA" small-block engine "go-fast" performance recipes, but up to now all development and "proof-testing" has been on 340- and 360-based combinations.* Everyone assumed that what was good for the large engines would more or less work on the 318, too. Besides, serious enthusiasts would always opt for more cubic inches, right? Not necessarily; at least not in 1988. While 10 years ago upgrading to a larger-displacement engine was relatively inexpensive, today finding and purchasing a 360 core in reasonable shape may be cost-prohibitive for many enthusiasts. After all, 360s today are used only in the larger pickups and vans. By contrast, the 318 has been in production since 1967, and there are an estimated five-million-plus engines out there in the "real world." Also, the majority of the "competition" is no longer 350- or 400-inch engines, but rather 305s and 302s.

Chrysler engineers, in conjunction with the dyno test experts at Arrow Racing, thought the time was ripe for an evaluation of the 318's street performance potential. Is it just a smaller 340/360? How much power could the dyno floggers wring out of the motor without sacrificing streetability?

PHASE 1

In an attempt to find some answers, Arrow began the tests with a 1986 nonroller cam 318 "police" engine with stock cast pistons, a cast crank, production rods, and late-model "high-swirl" mileage/emissions heads. To ensure the dyno mule's long-term survival, several durability upgrades were incorporated into the motor, as noted in the accompanying sidebar, "318 Short-Block Buildup". Initially, Arrow established baseline configuration using a mild production 318 cam (240 degrees advertised duration, .390-inch-intake/.400-inch-exhaust lift) and a stock (non-Lean Burn) electronic ignition. The 8:1 CR engine was dressed with a complete production exhaust system consisting of the stock cast-iron exhaust manifolds, catalytic converter, and mufflers. A production Thermoquad carb and complementary stock intake handled the fuel metering chores. Arrow retained the stock oil pan, but overfilled it one quart as a fail-safe measure. In this configuration, the engine produced 186 hp at 4000 rpm, about average for a late-model, low-compression,

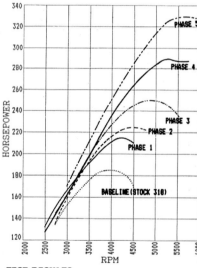

TEST RESULTS

All tests were conducted on Arrow Racing's computerized dyno, which corrected the results to standard sea-level atmospheric pressures and temperatures. Test numbers correspond to the text's numbered "recipe ideas."

smogged-out engine (see "Test Results" graph).

Next, Arrow began the hop-up process by installing a slightly hotter 360/2-barrel production cam (252 degrees advertised duration, .410 lift, 33 degrees overlap, 112-degree centerlines). Also added was the Mopar Performance electronic ignition con-

version package consisting of a vacuum-advance breakerless distributor with a performance-oriented advance curve, performance Orange Box control unit, and a wiring harness. For those who already have factory electronic distributors in good condition but are stuck with the weak "smog" brainbox, the Orange Box is also available separately in a kit with lightweight advance springs (P4007968). Arrow installed the new distributor using its "out-of-the-box" performance curve; total timing ended up at 35 degrees, with 10 degrees initial on the block.

Even though the 360 cam's paper specs were not much hotter than the 318 cam originally tested, when combined with the hotter ignition and performance advance curve, the little motor immediately recorded a big 31-hp increase, peaking out at 4250 rpm, where 217 total hp was produced. Just as important, the cam lost nothing on the bottom end, producing more usable power from 2750 rpm on up. This cam would probably not be detectable at any emissions check, and most likely would not significantly affect gas mileage, either.

ers. Improve the induction with a better intake and carburetor: A Holley 750 was chosen not because it's necessarily superior to the Thermoquad, but because the latter is out of production, making it hard to obtain replacement parts. Finally, add a stronger—yet still streetable—cam. Ground on 113-degree centerlines with .430-inch lift, 260 degrees advertised duration, and 34 degrees overlap, Mopar Performance's high-performance/RV grind offers good low- and midrange power. As with Mopar's other performance Purple Shaft cams, the bumpstick comes boxed with 16 new hydraulic lifters, break-in lube, and installation instructions.

These tried-and-true hot

rod techniques—exhaust, induction, camshaft—really awakened the 318. Horsepower was up throughout the curve over the previous 360 head test; and, the new coordinated combo outperformed the 318 combo over 3200 rpm, while hardly giving anything away below that point. Above 3500 rpm, the package left all its predecessors in the dust, climbing to its 251-hp peak at 5000 rpm, where it produced 26 more hp than the previous combo's best numbers. Even at the same rpm as Phase 2's power peak, the "triple play" combo was 21 hp stronger. It just goes to show what a good diet, deep-breathing exercises, and getting rid of the restrictive exhaust system can do for you!

PHASE 2
225 HP
2 stock 360 cylinder heads with stock springs, retainers, and keepers (1.88 intake/1.60 exhaust)

Perform performance valve job

PHASE 2

If a 360 cam worked so well, why not try "big brother's" cylinder heads? Compared to a 318, even run-of-the-mill 360 heads have larger valves and intake/exhaust ports. That's exactly what Arrow did next, bolting on a set of nondescript, used production 360 castings treated to nothing more than a good three-angle valve job. Overall, the bigger heads proved to be worth eight additional horsepower at a slightly higher 4500 rpm, and they were actually a tad weaker below 3500. Eight hp isn't a cost-effective mod when compared to the expense of purchasing a different set of heads.

PHASE 3
251 HP
2 stock 360 heads with stock retainers and keepers (1.88 intake/1.60 exhaust valves)
16 valvesprings, general purpose high-performance, P4120249
1 .430-lift hydraulic camshaft and tappet package, P4286669

1 Holley 0-3310 750-cfm vacuum secondary carb, P4349228
1 Edelbrock 2176 Performer intake manifold, P4286531
1 set of 1⅝ primary x 3-inch collector headers, P4286437; open exhaust

---

PHASE 4
290 HP
1 .455-lift hydraulic camshaft and tappet package, P4286671
8 2.02-inch intake valves, P3690230
8 1.60-inch exhaust valves, P3690231
16 chrome-moly valvespring retainers, general purpose high-performance (⅜ stem,

eight-degree keeper grooves), P4452033
2 sets hardened valve stem locks (⅜ stem, eight-degree, triple lock groove) (16 per set), P4120620
1 Gold Box competition ignition control unit, P4120600 (optional)

Port and polish cylinder heads

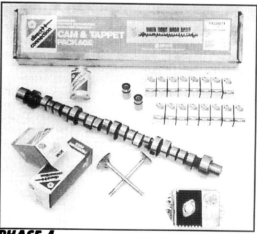

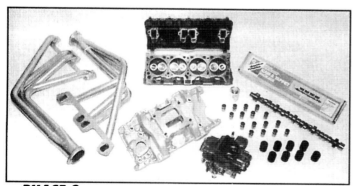

PHASE 3

The rather disappointing eight horsepower gain realized with the 360 head swap shows the necessity of assembling a coordinated package in which all components complement one another.

With the stock intake and smog exhaust system, the 318 "choked" under the 360 head's pressure. Solution: Relieve the choke points! Open up the exhaust by adding a good set of street head-

PHASE 4

Continuing with traditional hop-up methodology, Arrow next bumped up the compression, ported and polished the cylinder heads, and installed yet another larger (but still streetable) camshaft. In all previous tests, the compression had remained at a stock 8:1—low enough to operate on regular or unleaded gas. Now, the compression was raised to approximately 9:1 by milling the existing 360 heads .040 to .045 inch.

For a cam, Phase 4 added part No. P4286671. Intended

as a performance upgrade over the old obsolete 340 high-perf cam, the new offering features .455-inch lift, 272 degrees advertised duration, 48 degrees overlap, and 112-degree centerlines. Tappets and break-in lubricant are included in the package, but the recommended P4120249 valvesprings must be purchased individually (they were installed on the dyno engine during the previous test phase). Stock valvespring retainers can be used, but the aluminum or chrome-moly high-performance pieces are preferred.

A ported and polished "bracket 360" head is available from Chrysler, but it includes exotic racing valvesprings and titanium retainers that weren't needed with the relatively mild cams being run. Therefore, it is more cost-effective to add larger valves and port and polish the existing heads than to purchase the new, fully assembled and ported castings.

Finally, Arrow used the full-comp Gold Box ignition control; total overkill on this type of engine, but the company wanted to "run an ignition system test."

The Phase 4 modification program produced the largest power gain yet—39 more horsepower. No power was lost on the bottom, and from 3750 rpm up the new package left the previous combo in the dust. Power now peaked 500 rpm higher at 5250 rpm, where 290 hp was recorded; and the power curve remained virtually flat all the way through 5750 rpm. The peak 290 hp was about equal to the power level developed by "the competition" when treated to equivalent mods. Would this turn out to be the ultimate 318 street combo as well?

er 1.88-inch-intake/1.60-inch-exhaust valves in place of the stock 1.78-inch/1.50-inch combination. Even so, the intake valve was still smaller than the fully ported 360 head's 2.02-incher, which so far in testing had produced the best numbers. The 318's runners were also significantly smaller.

Arrow installed the ported heads on the 318 motor; the rest of the combo remained the same as the previous test. The results were staggering: Mopar's new ported 318 high-swirl head buried the larger 360 heads, producing an additional 41 hp at 5750 rpm! Even though the new peak was 500 rpm higher than Phase 4, the head lost nothing in the low- and midrange. In fact, the swirl head out-pulled all previous combinations from the bottom on up, and near the peak the power curve was virtually flat from 5250 through 5750 rpm. The new ported 318 heads worked so well that Chrysler is now offering them in fully assembled form.

This, then, was the ultimate combination: A 9:1 318 that produced 331 hp with only a .455-inch-lift cam, producing 1.12 horsepower per cubic inch! How many other equivalently sized and cammed packages have you seen that could match that kind of number? Not many.

What did we learn from this test? First, that different size engines from the same family do not necessarily respond in a like manner. Later tests of ported high-swirl 318 heads on a 360 would produce some gains, but nothing near what was accomplished on the 318 engine. Apparently, the heads are optimized for a 318-size engine. Second, cams considered relatively mild on a 360 produce respectable horsepower gains on a 318 without sacrificing low-end performance. Mopar Performance showed once again the importance of a coordinated package in maximizing performance potential. Finally, we come to the horsepower-per-dollar department: A 318 costs half of what a 360 costs. You can easily find and bolt together an early-'70s 318-powered A-body (Duster, Demon, Dart, Valient, etc.) for about $1000. The rest of the engine hardware will run between $2000 and $3000. Thus, for *under $5000* you can have a performance car that will blow late-model $15,000-plus production Detroit iron into the weeds. In fact, the boys at Mopar Performance are planning to do just that: They've purchased an old 318 Duster, and we plan to follow along as they shoot for mid-12-second e.t.'s. while still remaining fully streetable. Stay tuned.

PHASE 5

The Mopar boys had one last trick up their sleeves. As we've seen in various articles over the past several years, the current 1980s' trend in "race" wedge-style heads is toward a generic "heart-shaped" high-swirl combustion chamber, regardless of whose name is stamped on the valve covers. Among the competition, this design trend is treated as a great leap forward—a revelation that has only recently become apparent. Well, that's certainly not the case at Chrysler, where factory researchers began playing with heart-shaped chambers on the small-block

A-engines at least as early as the 1969 Indy program and 1970-'71 Trans Am effort. This early work lay dormant until the early '80s, when the factory incorporated a "swirl" design into the 2.2L 4-cylinder production heads. The chamber worked so well that in 1986 Chrysler developed a high-swirl 318 head, complete with recontoured intake and exhaust runners, and the resurrected heart-shaped combustion chamber.

Even though "swirl" had its roots in hard-core racing, the production 318 casting was optimized for low emissions and high gas mileage; improved power output was not a conscious factor in its design. This was borne out by the initial baseline and Phase 1 tests, which incorporated the high-swirl 318 head. As we've seen, 360 heads help to improve power production, which wasn't surprising due to the latter's larger valves and runners. But then it was decided to give the 318 heads another chance; they were ported using current high-swirl racing techniques, with a goal of maximum velocity rather than the old, outmoded "hog it out, square it up" volume routine. Also added were larg-

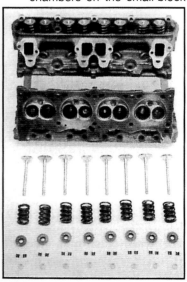

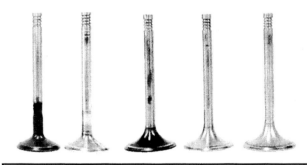

*Mopar Performance competition replacement valves were used in some of these tests. **From left,** the stock 360 1.60-inch exhaust valve is heavier and has inferior head shape compared to its performance replacement. The stock 360 1.88-inch intake valve flows less than its back-cut 1.88 replacement. The oversize 2.02-inch valve is also offered.*

318 SHORT-BLOCK BUILDUP

The bottom end of the Arrow Racing 318 remains surprisingly mild. What mods were performed ensured that the engine would survive long-term dyno service, and may not be necessary for the average, reasonably driven street engine.

MAJOR INTERNAL COMPONENTS

All part numbers with a "P" prefix are available from Mopar Performance/Direct Connection. Parts not listed are peculiar to each dyno package and are listed in the main article.

BLOCK	Stock 1986 "police," deburred and honed with torque plate
CRANK	Stock 1986 "police," checked for straightness
PISTON RINGS	Speed-Pro R-9229 +.005: 5/64 plasma-moly 1st, 5/64 cast-iron 2nd, 3/16 standard tension oil
MAIN BEARINGS	No. 1-2-4-5 positions: P3690655, babbitt, std.* No. 3 position: P3690656, babbitt, std.*
ROD BEARINGS	P4007026, babbitt, std.*
CONNECTING RODS	Production 318
ROD BOLTS AND NUTS	P4120097 set, 3/8-inch-diameter, high-strength steel
OIL PUMP	P4286589 hi-po assembly
OIL FILTER ADAPTER	P3690884 right-angle design (required to clear headers)
VIBRATION DAMPENER	P4452775 fluid design, 7¼-OD, degree'd
TIMING CHAIN	P4120262 roller timing chain and sprocket set
CAM KEYS	P4286500 offset key set (1-2-3-4-5 degree)
PUSHRODS	Production 318 (nonroller)
HEAD GASKET	P4120093 set, Fel-Pro Teflon-coated
VALVE COVERS	P4349675 set, aluminum
SPARK PLUG WIRES	P4120716 set, silicone insulation, metal-core
SPARK PLUGS	Champion N9Y

* Sold individually.

BUILD SHEET

Blueprint specs listed below are for final Phase 5 high-swirl configuration. For data not listed below, refer to stock service manual. All dimensions in inches or fractions thereof, unless otherwise stated.

ENGINE BUILDER: ARROW RACING ENGINES, 3811 Industrial Dr., Rochester Hills, MI 48057, 313/852-5151 (address and phone number effective 3/15/88).

DISPLACEMENT	318
BORE X STROKE	3.91x3.31
COMPRESSION RATIO	9.2:1
PISTON DECK HEIGHT	.066 below
CYLINDER HEAD VOLUME	57cc
PISTON-TO-VALVE	(@ 0 lash)
Automatic trans	Intake: .090 Exhaust: .100
Manual trans	Intake: .100 Exhaust: .110
PISTON-TO-HEAD (w/gasket)	.055 (minimum) .120 (actual)
PISTON-TO-WALL	.003
PISTON RING END GAP	1st: .015 2nd: .012
ROD/MAIN BEARINGS	.0025
CRANK END-PLAY	.004
ROD SIDE CLEARANCE	.012
CAM (P4286671)	112-degree intake centerline, installed @ recommended 108-degree centerline
VALVESPRING HEIGHT	1.650 to 1.670
CAMSHAFT END PLAY	.005*
HONE PROCEDURE (CK-10)	Use 625 stone with torque plates installed on both cylinder heads (if only one plate is available, install cylinder head on opposite deck)
ROD BOLT TORQUE	45 lbs.-ft. (oil)
HEAD BOLT TORQUE	105 lbs.-ft. (oil)
SPARK PLUG GAP	.040
TIMING	35 degrees total @ 3000 rpm

* Chrysler A-engines use a camshaft retaining plate. This figure is the required cam-to-plate clearance.

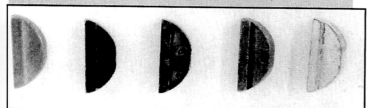

The offset cam key set permitted dialing-in each cam tested "straight up."

Compared to a standard head (bottom), the new high-swirl 318 head's exhaust ports (top) have a higher roof. The bottom of siamese center ports has also been filled in.

Pushrod holes (arrows) on the high-swirl head (left) are enlarged to provide proper clearance in production hydraulic roller cam applications. This prevents any application of the old, "hog it out" school of cylinder head porting, at least on the intake side.

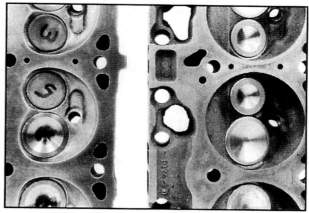

The late 318 head's heart-shaped combustion chamber (left) helps create swirl—current race cylinder head technology includes chambers such as these. Piston to-head clearance should be checked if using 11:1 or higher-compression ratio pistons.

A
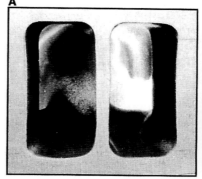

B
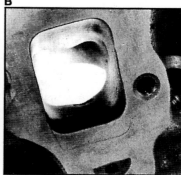

Intake (A) and exhaust (B) ports on ported high-swirl heads are contoured to maintain maximum velocity and swirl effect. In this case, bigger isn't necessarily better. **HR**

350 CID SMALL-BLOCK MOPAR

POWER EXPLOSION!

BY JIM LOSEE

A Power Injection For Mopar's 318

We bet you didn't know that Chrysler made a 350-cubic inch small-block. A 350 Chevy, yes, but a 350 Mopar? Actually, Chrysler never did make an LA (small-block) engine with 350 cubes, but they certainly have built thousands of the 318 engines since 1967. The 318 has a fine durability record any manufacturer would be proud of, but it lacked the muscle of the 340 of yore basically due to its lack of cubic inches.

We discussed the notion of adding some additional cubes and muscle to a 318 with Alan Welch and Larry Revis of Automotive Balancing Service and with Rich Baumann and Jeff Ginter of Speed-O-

Motive. Our most important criteria was avoiding an expensive proposition for the average street enthusiast and the ability to develop this engine with readily available parts requiring little or no custom machining.

The 318 uses a 3.31-inch stroke

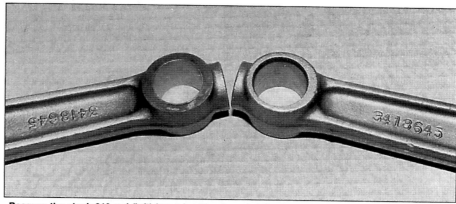

Because the stock 318 rod (left) has a bushing for a full-floating wristpin and the cost to re-bush these rods is $100 a set, the stock 360 pressed-pin rod (right) is used in the Speed-O-Motive stroker kit.

crank. Because of the low cost and the ease of access, it was decided that a 360 crank with a 3.58-inch stroke would be the ticket for added muscle. The 360 main journals and rear seal area have to be ground down to the stock 318 size crankshaft, while the rod journals between the 318 and 360 are the same diameter and needed only to be polished. This required a minimal amount of work for the expert crank grinders over at Castillo's Crank Service who also chamfered and radiused the oil holes and made sure the crank was straight.

With the crank problem solved, next came the dilemma of finding a piston to fit the 318 bore and adapt to the 360 stroke. We analyzed considerations such as pin diameter and compression distance, along with adapting the Chrysler 318's full-floating variety wristpins. With these thoughts in mind, the A.B.S. and Speed-O-' gang came up with a Speed-Pro piston from a 318 engine overbored 0.030 inch. This piston is used on '67-69 and late-'73-86 318 engines and has a flat top with two sets of double eyebrow valve reliefs. The piston rings required the use of 1/16-, 1/16-, 3/16-inch-thick grooves and have moly composition. Using this piston means that the 318 block is bored 0.030-inch oversize for a bore diameter of 3.940 inches.

The next problem we encountered was how to secure the wristpin in the connecting rod. The standard 318 rod utilizes a bushing for a full-floating pin and the piston uses wire locks to hold the pin in place. The 360 rod uses a pressed pin to secure the piston to the rod. We "side stepped" the $100 machine shop charge to rebush the rods by simply press-fitting the 360 wristpin in the 318 rods.

Next, we considered the dome

A.B.S. cuts the piston top 0.100 inch to allow the 318 piston to be used with the 360 crank. This puts the piston 0.020 inch down in the bore and helps yield a compression ratio of 8.8:1.

Allowing the 360 crank to fit into the main bearing saddles of a 318 is accomplished by cutting down the main journals. When Castillo Crank Service cut the mains down they left a generous fillet, or radius, retaining the strength of the stock 360 crank.

Because the rear seal area of the 360 crank is larger than a 318, it too has to be turned down. The key to this operation is getting the height of the seal flange right, and Castillo Crank Service came through.

Having adequate lubrication and a scratch-free surface for the crank to ride on is paramount to engine longevity. Workmanship done by Castillo Crank Service, both in micro-polishing and chamfering the crank journals is flawless.

Speed-Pro main and rod bearings were used for Speed-O-Motive's 350 Mopar Muscle engine. Clearances were set at 0.002 inch for the rod bearings and 0.003 inch for the main bearings. Thrust clearance was 0.008 inch with rod side clearance at 0.012 inch.

350 CID SMALL-BLOCK MOPAR

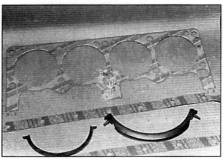

Sealing the engine is done via a full Fel-Pro gasket set that includes a 0.040-inch-thick head gasket and cork oil pan rail gaskets. A dab of RTV in the corners between the end seals and rail gaskets prevents leaks when the oil pan is installed.

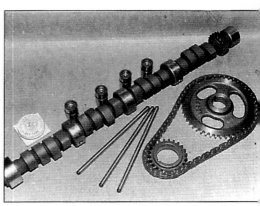

Actuating the valves is done via an Isky 260 Supercam. With a mild lift of 0.435 inch and only 208 degrees at 0.050 inch, this makes the 350 incher run hard.

height, which added to the overall compression distance. Larry Revis milled 0.100 inch from the dome and brought it within Chrysler specs to acheive a reasonable 8.8:1 compression with a 0.040-inch-thick compressed Fel-Pro head gasket with the piston sitting 0.020 inch down in the bore. Even after these modifications, there was still over 0.265 inch of material left in the piston dome, well above the required amount for safe engine operation even with nitrous.

To ensure reliability, the reciprocating assembly was dynamically balanced at A.B.S., including the torque converter from A-1 Automatic Transmissions. This is done because all 360s are externally balanced and Chrysler considers the torque converter part of the rotating assembly. A complete Speed-Pro bottom end was installed to factory specs, again to gain the most in reliability. This included bearings, high-volume oil pump, and moly piston rings. To seal everything up, a set of Fel-Pro gaskets were used along with a little RTV.

For valve actuation with streetability the main concern, an Isky cam of 260 degrees advertised duration and 208 degrees at 0.050-inch lift was installed. Lift for this cam is pegged at a mild 0.435 inch, while lobe centers are at 108 degrees. Isky valvetrain components were used, which included chrome moly pushrods, steel timing chain and gears, and assembly lube for the cam. A mild cam such as this guarantees that the idle will be smooth and the engine will breathe at the top end.

Cylinder heads were the next area of concern. One of the best-prepared set of LA Mopar heads comes directly from the shelves of Mopar Performance. Our castings started life as 360-cube truck parts and are used because of their free-flowing intake

An Edelbrock Performer intake manifold (PN 2176), is used due to its strengths in helping to make both low-end torque and higher-rpm power. Sealing the manifold to the head is done using a Fel-Pro gasket and RTV.

A high-volume Speed-Pro oil pump is used to circulate lubrication throughout the engine. When screwing the pick-up into the pump body, be sure to use Loctite 271 on the threads so the pick-up doesn't vibrate loose.

When bolting down the Speed Pro high-volume oil pump make sure that the sleeve for the oil pump driveshaft fits down into the main cap. If this isn't done correctly, the oil pump body will break.

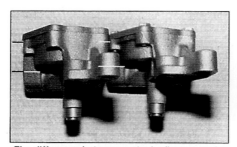

The difference between a stock oil pump and a high-volume unit is readily apparent in this photo. The high-volume pump on the left has a larger body and taller gears, allowing it to circulate more oil.

If your 318 has more than 100,000 miles on it, replace the distributor shaft bushing at the rear of the block. If this bushing is worn out it can cause erratic ignition timing and can lead to oil pump failure.

350 CID SMALL-BLOCK MOPAR

and exhaust ports. These heads, known as the "bracket heads," carry part number P4286864 and come fully ported and cc'd with heavy-duty valvesprings and titanium retainers. For our application, the heavy-duty valvesprings were a bit too strong and might have caused premature cam wear, so they were replaced with some milder Isky springs. Other than that, these heads were used as is and bolted on using stock valvetrain components and a Fel-Pro Perma-Torque blue head gasket.

On top of the heads, an Edelbrock Performer aluminum intake manifold was used to transfer the fuel/air mixture from the Holley 3310-2 750-cfm carb. This manifold and carb combination gives good low-end torque along with a significant increase in power at the upper rpm scale, while maintaining streetability. Ignition of the fuel/air mixture is handled via a Mallory Unilite distributor and a Hy-Fire amplifier and ProMaster coil. Sparking the plugs is done through 8mm solid-core silicone jacketed wires with straight ends. Hedman 1.625-inch-diameter tube headers are used to keep the velocity up and help evacuate spent exhaust gases from the combustion chamber.

With all these problems worked out using relatively little expensive machine work and utilizing off-the-shelf parts, the engine size isn't right at 350 cubes, but is close enough at exactly 349.19 cubic inches.

This combination is easy to assemble by using the Mopar Muscle kit from Speed-O-Motive, it's not expensive, and will resurrect a poor-running 318. With the small-lift cam and basically stock running gear, this engine is a model of reliability and will give good service, both on the street and track. Due to some time constraints, we weren't able to put the engine on the dyno, but when in-

After the cam is installed in the block and the timing chain and gears are about to go on, be sure to line up the timing marks properly. The cam gear mark is vague on this particular gear, so we had to look closely when lining it up with crank gear.

After the piston top was cut down 0.100 inch, it sits in the bore at top dead center 0.020 inch down below the deck surface. This helps the flame travel across the piston top and aid power production.

stalled in a '69 Plymouth Satellite, it turned into a ground-pounder compared to the stock 318. It goes to prove that a performance injection can really make some big-time Mopar Muscle. G

To ensure we had enough dome material left after cutting the top, the thickness was measured on this trick device built by the guys at Automotive Balancing Service. Even after the 0.100-inch cut our dome still had a 0.265-inch-thick top, enough even for a nitrous-equipped engine.

Mopar Performance supplied the heads we're using on this 350-cube muscle motor. They're known as the "bracket heads" and are derived from 360 truck castings. The heads come completely ported and polished with cc'd chambers and heavy-duty valvesprings and titanium retainers.

Equipped with a 2.01-inch-diameter intake valve and a 1.60-inch-diameter exhaust valve and a 65.5cc combustion chamber, these Mopar performance heads will help generate some substantial power from our 350 cuber.

SOURCES

A-1 Automatic Transmissions
Dept. CC
7359 Canoga Ave.
Canoga Park, CA 91303
818/884-6222

Automotive Balancing Service
Dept. CC
P.O. Box 1984
South Gate, CA 90280
213/564-6846

Castillo Crankshaft Repair
Dept. CC
14654½ East Firestone Blvd.
La Mirada, CA 90638
714/523-0321

Edelbrock Corp.
Dept. CC
2700 California St.
P.O. Box 2936
Torrance, CA 90509-2936
213/781-2222

Fel-Pro Inc.
Dept. CC
7450 N. McCormick Blvd.
Skokie, IL 60076
312/674-7700

Hedman Hedders
Dept. CC
9599 West Jefferson Blvd.
Culver City, CA 90230
213/839-7581

Holley Replacement Parts Div.
Dept. CC
11955 East Nine Mile Rd.
Warren, MI 48089-2003
313/497-4250

Ed Iskenderian Racing Cams
Dept. CC
16020 So. Broadway
Gardena, CA 90248
213/770-0930

Speed-O-Motive
Dept. CC
P.O. Box 4308
Santa Fe Springs, CA 90670
213/945-2758

Speed-Pro
Dept. CC
100 Terrace Plaza
Muskegon, MI 49443
616/724-5011

Finally, after 15 years of fruitless rumors, faithful Dodge and Plymouth fans can frolic in fact. A decade and a half down a now twisty road, a brand-new old Chrysler V8 has surfaced again. And all those tantalizing tales of a mythical Mopar muscle motor—the legendary Ball-Stud Hemi—can now be confirmed.

Unfortunately, what you're looking at here is now truly one-of-a-kind. Like many of the corporation's production performance motors, Chrysler's Ball-Stud Hemi engine succumbed to bad timing in the early Seventies. A combination of conditions including emissions and economy considerations, underwriters investigations, and mounting corporate financial pressures, made what could have become America's number one boulevard brawler into a fatal victim of circumstance—instead of King of the Street.

During the late Sixties, Chrysler's 426 Street Hemi had already claimed supremacy on America's avenues. But the big, heavy 8-barrel engine's relatively limited production volume (approximately 9000 were fitted to Dodges and Plymouths between 1966 and 1972) made the motor expensive both to produce and to buy. At the same time, the corporation's mass-produced big-inch engines (the 383/400 and the 440) were uneconomically based on two completely different short-block assemblies. Chrysler considered the Ball-Stud Hemi as an attempt to correct all of these limiting conditions.

But because the engineering efforts behind the ball-stud-style rocker arm engine were relatively short-lived, the project got minimal exposure. Few folks even knew much about the motor back then. And now, 18 years later, even less seems to be remembered.

HOT ROD recently stumbled on what is strongly suspected to be the sole surviving sample of the Ball-Stud Hemi at Dick Landy Industries (19743 Bahama St., Northridge, CA 91324, 818/341-4143). Long-time Mopar flogger Landy has a considerable collection of Dodge and Plymouth parts that includes some pretty peculiar pieces for the old V8's. But this all-iron Chevy Rat motor-style assembly has got to be his oddest hardware ever.

Similar in approach and in external appearance to the big-block Stovebolt scheme, Chrysler's Ball-Stud Hemi plan was to build a pair of high-volume, low-cost, high-performance powerplants based on a common low-deck cylinder block. Their proposed 400 and 440-cubic-inchers would share many of the characteristics of the ball-stud rocker-equipped stagger-valve, semi-hemi Chevys and Fords, including their limitations. In addition, the BS Hemi's redesigned rocker arrangements would trim both weight and costs, and they'd help make the engine fit better in a wider range of cars.

MOPAR MYSTERY MOTOR

THE BALL-STUD HEMI

Text and Photography: Al Kirschenbaum

Intended for assembly-line production somewhere between 1971 and 1973, the Ball-Stud Hemi was tentatively scheduled to replace three different blocks and two distinct cylinder head designs with one common casting of each. Chrysler went with the hemispherical combustion chamber approach primarily out of corporate tradition, but the built-in plusses of a high surface-area-to-volume ratio, big valves, and crossflow chambers made that design decision an easy one.

In backtracking the data on this engine, we also learned that some designers wanted to revise both motors' displacement slightly. A few extra cubes for the big-incher would, they reasoned, make the "444" designation an image-enhancer as well as an improved marketing tool. And bumping the smaller motor to 400 inches from 396 made sense for rather obvious reasons. Without exception, everyone emphasized that these were, without question, production powerplants, rather than any sort of race engines.

At first, this latest Chrysler V8 barely ran as well as their existing wedges. But in time, it turned out to be better than the A134 4-barrel 440 and not quite as good as the A102 8-barrel Street Hemi. From start to finish, there was maybe a year of development work involved in the project, with an emphasis on the heads, port design, and performance refinements. The best estimates indicate that there may have been as many as a dozen BSH engines built, or possibly as few as three. Although it never got into the endurance phases of testing, dyno development did become promising. And at least one Ball-Stud motor was reportedly fitted to an engineering mule (a '69 B-body Plymouth automatic).

Again, a variety of factors combined, in late '69, to grind the project to a halt. Aside from stricter emissions requirements and their negative effects on driveability, a general de-emphasis on performance made continuing development work difficult. And because quite a bit of new and different machinery would have been required to produce the Ball-Stud Hemi, costs were also escalating. This was at a time when the corporation was experiencing the severe financial

fective with large valve sizes, the canted valve design also allows the domed, or an almost-a-hemi (in this case) combustion chamber. Efforts to squeeze in a centrally located spark plug resulted in the use of 14mm, rather than 18mm, plugs.

Other departures from the traditional wedge head layout included a pair of lube drain-back passages at each end of the casting to channel returning oil to the pan. There are also head bolts at the ends of each bank for improved clamping. In addition, a smaller hex on the external head bolts provides additional clearance for wrenches.

stresses that almost crippled the company during the next decade.

Thanks to Dick Landy and the crew at DLI, HOT ROD was able to stand by while the world's one-and-only Ball-Stud Hemi was partially disassembled. During the process, Dick was quick to point out that while most of the experimental extra-duty parts developed by the Corporation for performance duty usually looked like handmade, one-off developmental hardware, this engine appeared to be a finished piece.

Every last component in the Ball-Stud combination looked like production was right around the corner. But it just goes to show you how quick the auto industry can take a turn for the worse. The photos that follow reveal some of the thinking and tinkering that went into what could have been one of the most memorable milestone musclemotors of all time—Chrysler's Ball-Stud Hemi.

CYLINDER HEADS

Bolted to the top of the cylinder head on partially ball-shaped threaded studs, the engine's rocker arm setup was both a radical departure from other Chrysler V8 designs as well as the source of the

engine's development name. In addition to its corporate A279 engineering designation, the project was known only as the Ball-Stud Hemi. Unofficially, we're told the engine was also affectionately referred to as the BS Hemi.

Because a preliminary engineering decision had been made to retain the standard B-engine (383/400/440-cid) head bolt pattern, the design of the Ball-Stud Hemi head's exhaust ports was badly hampered. Although the head wanted a direct flow path (like the 426 Hemi), the exhaust port used an S-shaped twist to sneak around some outboard head bolts. As it turned out, the head bolt pattern was revised slightly toward the end of the project to try to straighten out the port.

Unfortunately, however, the early design plan had already dictated some of the heads' other tooling, and out-bound airflow remained considerably restricted. This resulted in a chamber arrangement that's not really a true Hemi (where the valves are laterally opposed, rather than slightly offset as they are here). Compared to the Hemi, port area was greater on the intake side (3.575 square-inches) and less (2.488 square inches) in the exhaust port. The engineers involved felt that had they been able to relocate more of the cylinder head fasteners, they could have had a ball-stud rocker engine that was every bit as good as the 426 motor.

In contrast to an in-line valve engine, the Ball-Stud Hemi's intake valves are located closest to the intake manifold side of the head, while the exhaust valves are positioned transversely on the opposite, or exhaust manifold side of the casting. This staggered setup orients the canted valves at compound angles to the intake and exhaust ports, and enhances airflow by reducing valve shrouding and eliminating the sharp runner turns involved in in-line valve layouts.

This "polyangle" arrangement also featured equally spaced intake ports (like the Ford 429SCJ, rather than the Chevy Rat motor) for more consistent mixture distribution than in the traditional wedge head port layout. Especially ef-

ROCKER GEAR

In the ball-stud rocker pivot setup, the sides of the U-shaped "bathtub" rockers helped maintain alignment, while the fulcrum/stud had a positive-locking shoulder. As in the variety of competitors' motors, the lock nut was simply torqued down to set the valve lash. In production, the heat-treated, ground shoulder would have to be held to close tolerances to stay within the non-adjustable hydraulic tappet's range.

The size of the rockers was, in part,

another compromise dictated by the chamber, valve, and pushrod layout that were, in turn, concessions to the original plan to retain the B-wedge head bolt pattern. Although it checks out at an effective 1.58:1, the rocker ratio is 1.6:1. The

severe pushrod angles and the resulting rocker-train geometry causes a small loss of lift.

Lube for the rockers was supplied via the lifters through hollow pushrods. Guide plates were used to hold and align the pushrods with the rocker arms and to guide the rockers in travel to match the angle of their valves. Chrysler did design one other ball-stud rocker arm engine, but the 6-cylinder 245 Hemi was built only in Australia for use in that country.

VALVES

Sized to match the 426 motor (2.25-inch intake, 1.94-inch exhaust), the valves were situated 21 degrees apart (intake canted 15 degrees and exhaust canted 6 degrees from centerline of bore). These locations were the result of

idealistic attempts to position the intakes and exhausts directly opposite each other, like the existing 426. Because this wasn't possible, and because the valve seats had to mate with a hemispherically shaped chamber, the valves also ended up canted, rather than positioned square

to each other. For comparison, a pair of 426 valves are illustrated here to the right of the BSH valves in the middle, while a set of standard 2.08-inch B-wedge intakes and 1.74-inch exhausts are on the left. Multi-groove locks promote exhaust valve rotation.

BLOCK

With the specific structures required by its unique low deck and the extra oil drain-back passages, head bolts, and core plugs, the BS Hemi cylinder block was considered an all-new casting. And although the blocks were considerably different than both of their wedge coun-

terparts as well as the Hemi, a few common features (like the 426/440 journal sizes) were retained. The basic blocks were similar enough for the earliest prototype Ball-Stud motor to use a modified version of an existing wedge casting, but because the block was basically a B-wedge, going to 4-bolt or cross-bolted maincaps was not a practical or cost-effective proposition.

Both blocks' large bores made it especially important to have coolant in the water jackets rather than casting sand. This concern over core clean-out led to an additional pair of core holes located at the rear face of the block and another hole at the left front. Clean-out in the BS Hemi was even more critical because of its cast-in oil passages at both ends of the casting.

OILING

Because a ball-stud rocker arm system requires a considerable volume of lubricant, concerns over oil drain-back from that area prompted an extra draw-down passage at each end of the heads and block. There were also plans to evaluate the standard-size ⅜-inch oil pump pick-up tube in the endurance testing phases of development, but work on the engine never progressed that far. For reference, the 426 Hemi's sucker pipe measured ½-inch diameter.

PISTONS

Permanent mold cast-aluminum slipper-skirt pistons with an offset pin and autothermic expansion control were designed to permit a close bore fit (less than 0.002-inch) and provide good durability along with excellent oil and compression control. Installed, the piston tops were measured and found to be approximately 0.180-inch above the block deck. Even with relatively deep valve reliefs, DLI's measurement of displaced block and chamber volume indicated a 9.8:1 compression ratio. From what we learned from engineers who were involved in the BSH project, the target ra-

tio for production was more like 10.5:1.

With a moly-filled top set, piston ring design appeared to be consistent with the high-performance 440 wedge engines of the era. The pistons even had a pair of standard-style orientation notches on top to indicate their direction (towards the front) of installation. The unique snap rings shown here were evaluated because of the 426 Hemi's early tendency to pound-out the pins at high rpm. From what we determined, these unusual retainers may have been just one of the many types being tried at the time.

WRIST PINS

The tight fit (less than 0.0005-inch) and the generous wall thickness of the Ball-Stud motor's full-floating wrist pins may have been dictated by the production-style pistons used. Cast pistons expand at a different rate than forged slugs, and thick-wall pins will also help to support the "flexible" split-skirt piston.

CONNECTING RODS

In order for the same low deck block casting to suit both displacement configurations, the long-stroke version of the Ball Stud motor used the forged con rod

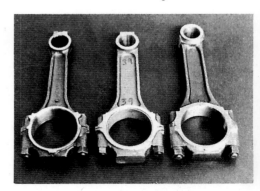

shown here in the middle (between a standard 383/440 rod on the left and an OE 426 Hemi rod on the right). The short

ENGINE	BORE/STROKE	PIN SIZE	ROD LENGTH	VALVE SIZES		JOURNALS		PORT AREA*	
				INTAKE	EXHAUST	MAIN	ROD	INTAKE	EXHAUST
400W	4.34 x 3.38	1.09	6.385	2.08	1.74	2.625	2.375	2.537	1.840
440W	4.32 x 3.75	1.09	6.768	2.08	1.74	2.750	2.375	2.537	1.840
426H	4.25 x 3.75	1.03	6.860	2.25	1.94	2.750	2.375	3.325	2.720
400BSH	4.32 x 3.38	1.09	6.385	2.25	1.94	2.750	2.375	3.575	2.488
440BSH	4.32 x 3.75	1.09	6.385	2.25	1.94	2.750	2.375	3.575	2.488

ENGINE SPECIFICATIONS

* Note: 400W and 440W cylinder head measured was a 2843906 casting.

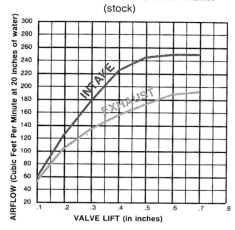

CHRYSLER BALL-STUD HEMI
(stock)

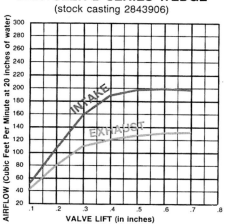

CHRYSLER B-SERIES WEDGE
(stock casting 2843906)

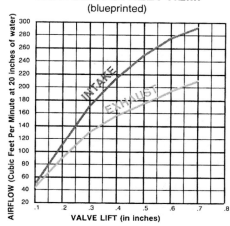

CHRYSLER STREET HEMI
(blueprinted)

forged rods were dictated by the lower block height, the location of the piston pin (it couldn't be raised higher without running into the bottom ring), and on the other end, the crank counterweights that were already close to the bottom of the piston skirt.

To retain high-quality metallurgy, the Ball-Stud rods' big ends were finish-ground rather than bored and finish-honed. Landy pointed out that although the short BSH rod used wedge-style ⅜-inch hardware (as opposed to the Hemi's ⁷⁄₁₆-inch nuts and bolts), the forging would have made an excellent basis for a performance rod for all of Chrysler's B-based V8's.

CRANKSHAFT AND BEARINGS ■
With the exception of the length of their rod throws, crankshafts for both versions of the BSH were similarly designed. The large-displacement assembly at DLI had a standard 440 forging that was externally balanced with an off-

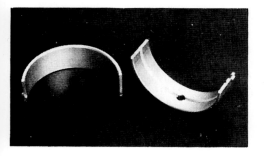

set weighted damper that helps compensate for the odd rod and piston masses.

In a departure from the standard wedge assemblies' Clevite 77 inserts, the Ball-Stud motor used aluminum main bearings (for increased load durability) that were grooved in their top halves only. The alloy inserts are about 20-percent tougher than the standard tri-metal bearings, but they're also more expensive. And because of their poor imbedability, they're not as forgiving as the slightly softer Clevites. Like most of Chrysler's production V8 powerplants, Clevite 77 connecting rod inserts were also used in the Ball-Stud assembly.

INTAKE MANIFOLDING ■
Another area that was never fully resolved, the BS Hemi's intake manifolding may have ended up considerably different than the hardware shown here. In addition to the dual-plane spread-bore iron casting that was fitted to this particular version of the engine, another dual-

level design was in the works that fed each bank of cylinders from separate levels of the manifold. Increasingly stricter emissions requirements prevented a dual 4-barrel system from even being considered for the Ball-Stud combinations, but there was talk of using a considerably higher-flow Carter Thermo Quad than had been used before.

EXHAUST MANIFOLDING ■
Designed with the same port spacing as the 426 Hemi, the header-like exhaust manifold castings were the result of compromises based mainly on assembly line body-drop requirements. Al-

though they were found to be reasonably efficient, they were nowhere near optimum. And although the development program was over before the manifolding was sorted out, testing indicated that there was a power loss attributed to the castings being tried.

Imagine what hot rodders might have done with this engine if it had been produced! **HR**

DART GAMES

By Todd Howard

Building the Chrysler 340 A-Engine for Our Grand Prize '69 Dodge Dart Swinger

We used Perfect Circle engine parts and Victor gaskets almost exclusively throughout the build-up. Note that all components are factory replacement type items, right down to the cast pistons and stock cam.

This piston/valve combo is representative of what you can expect if your aluminum valve spring retainer breaks. Inspection of the original engine revealed four more retainers that were ready to give way.

Bob Lambeck was in charge of machining and building the 340 powerplant to grace the engine compartment of our giveaway Swinger. Block prep operations included boring, deck plate honing, line honing, and cam bore honing.

Now that St. Louis International Raceway has been designated as the official location for this year's CAR CRAFT Street Machine Nationals, we are hard at work on the official Grand Prize street machine. As the star attraction of the Dart Games series, our '69 Dodge Dart Swinger will soon be the property of some lucky Street Machine Nationals participant. And since the number of entries we are accepting this year is reduced to 4000, you now have a better chance at winning this restored factory musclecar.

After rebuilding the transmission and rearend last month, we thought the next logical step would be to move on to the powerplant. For those of you who were with us for the introduction of this project, you know about the unfortunate mishap that befell the original 340 that came in the car. It succumbed to a broken aluminum valvespring retainer which deviously allowed the Number Seven intake valve to punch a hole in the piston. The motor, however, was still salvageable.

Originally, the 275 hp 340 engine was outfitted with 10.5:1 pistons, Carter AVS carb, dual-plane intake, forged steel crank, 2.02-intake, and 1.60-inch exhaust valves, full-floating wrist pins, and high-flow exhaust manifolds. While it wasn't Chrysler's biggest horsepower option in '69, it was the stoutest A-block available.

To rebuild the engine, we contacted one of the nation's leading Mopar Stock Eliminator racers, Bob Lambeck. Bob has a small, yet well-equipped, machine shop in Van Nuys, California, and was more than willing to help us extract the maximum amount of horsepower out of a stock rebuild.

While quality workmanship is a major consideration when building a good street performance engine, you first need a supply of good parts to work with. Perfect Circle and its sister company, Victor, came to the rescue with practically every engine component and gasket required for the build-up. Of course there were a few parts we still had to obtain, including a stock single-point distributor, an original 4-barrel intake manifold from 'Cuda Connection, a new Carter AVS #4639S service replacement carb, and an original oil pan. These were the components changed on the engine when we received it from Tom Toycen.

With the Perfect Circle parts, Bob

DART GAMES

was able to assemble the block with close tolerances for maximum power. Machining operations included line honing the main journals, boring the block .030-inch over, torque plate honing the cylinders, machining the crank, sizing the rods, and installing bronze-wall valve guides.

These are the standard machining techniques applied to every engine that goes through Bob's shop, Mopar or not. But since this was a small-block Chrysler, Bob had an additional list of items to check. For one, he measured the cam bearing bores for size and honed when necessary. And because the 340 employs non-adjustable rocker arms, Bob was careful to maintain uniform valve depth to prevent drastic variations in the valvetrain geometry. Bob has found through experience that Chryslers are also prone to deck alignment variations and proceeded to cut .005-inch from each cylinder bank as measured off the crankshaft. The amount of metal removed from the block deck is critical here since it ultimately affects the height of the piston in the bore (piston-to-deck height). This operation also has an effect on the correct amount of lifter preload.

In addition to decking the block, Bob shaved .005-inch off each head surface for maximum gasket sealing and to reduce the size of the combustion chambers. As a general rule for Mopar A-engines, a .0048-inch cut reduces the combustion chamber by 1cc. However, for every .010-inch removed from the head surface, .012-inch of material will need to be trimmed from the intake surface of the head to maintain proper intake manifold alignment.

One very important point that deserves mentioning is the fact that Chrysler's advertised compression ratio for the '69 340 and the official engine specifications differ drastically. Given a combustion chamber volume of 63cc, a flat top piston with a positive "out-of-the-block" .045-inch deck height and a .020-inch thick steel shim head gasket, this should yield a 10.5:1 compression ratio. However, a simple calculation of these figures produces a figure more like 13:1! Hardly what you'd call acceptable for a street car.

On a more realistic scale, we found that by shaving both the block and heads .005-inch our engine developed

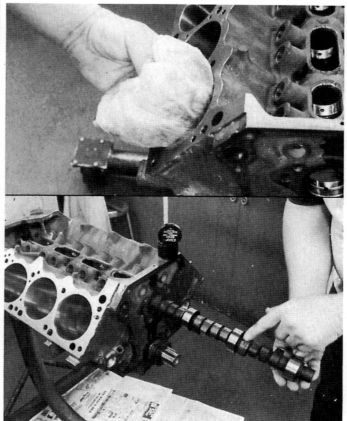

While most engine builders feel that washing a block with soap and water is sufficient before assembly, Bob has found that the honing marks in the cylinder bores can retain a small amount of grit afterward. Multiple passes with lacquer thinner-soaked paper towels is the only way to obtain a clean bore.

Perfect Circle offers a true replacement camshaft for hi-po 340 (PN 229-1654) that features a factory dual-pattern design. Duration measures out to 268 degrees intake/276 exhaust, while lobe lift is .430-inch intake/.445-inch exhaust.

DART GAMES

a healthy 10.4:1 compression, closer to the publicly-released figure. This was obtained by using a .044-inch thick Victor composition head gasket and Perfect Circle flat top pistons while maintaining .013-inch to .015-inch positive piston-to-deck clearance and 68.5cc to 69.1cc combustion chambers.

Other than this idiosyncrasy, assembly of our small-block Mopar is similar to most street engine rebuilds. Bob was able to point out several other tips that he has picked up while working with the A-blocks. For example, when installing the crankshaft, you should torque the four outer main caps first, leaving the Number Three main cap loose. Then force the crank forward

and torque the cap to specs. This helps to align the cap with the crankshaft, preventing potential binding problems.

Bob's valvetrain assembly tips include assuring that the lifters have a minimum amount of preload once the rocker shafts are torqued in place. Since the Chrysler 340 uses non-adjustable rocker arms (with the exception of the '70 Six-pack 340) valvetrain geometry is critical. It becomes especially important to check the preload when the decks and heads have been milled. If excessive material is removed, this can prevent the valves from closing due to the limitations of the hydraulic lifter. One way to solve this problem is to shim the rocker shaft above the rocker shaft stands. Experiment with the amount of shim stock required for your application. If this tech-

nique is used, be sure to use shim stock that allows engine oil up to the rocker shafts.

While Bob was finishing the final assembly of the 340, we took the stock distributor over to Bob Jennings Dyno Tune where Nick Jennings dialed-in a factory spec ignition curve. With the ignition functioning properly, we were off to 'Cuda Connection, the folks that are restoring our giveaway Swinger, where Randy Gerstenberg proceeded to coat the 340 with a layer of Hemi orange acrylic enamel. Completing our scheduled rounds was a trip to Edelbrock where Louie Hammel broke in the powerplant and put it to the test on their dyno. As you can see from the at-

The notches on the top of all stock and stock replacement pistons indicate the forward side of the piston as installed in the engine. This is because the piston pin centerline was ground offset to reduce piston slap.

Bob welded a baffle in our stock pan to maintain as much oil around the oil pump pickup as possible. This, combined with the stock windage tray, should take care of all our oiling needs.

A minimum amount of material was shaved from the heads to keep final compression in line. This resulted in combustion chamber sizes ranging between 68.5 and 69.1 cc's. Note the three angle valve job as well.

All Chrysler small-block rocker arms are not created equal. Identified as the left and right rocker arms, the difference lies in the pushrod offset and not whether the rocker is for an intake or exhaust valve.

LAYING IT ON THE LINE

While Chrysler originally rated the '69 340 at a conservative 275 horsepower, NHRA was quick to re-factor the stock powerplant for drag racing at 300 ponies. As you can see from the attached dyno chart, our Bob Lambeck-rebuilt 340 developed more power than NHRA's factor. With Louie Hammel at the helm of Edelbrock's Super Flow SF-901 dyno, our giveaway Swinger's motor churned out a healthy 314.7 peak hp and 361.2 lbs./ft. peak torque. Used in conjunction with the test were 1⅝-inch headers, a late-model factory electronic distributor, and 10 degrees initial and 36 degrees total ignition advance.

RPM	Corr. Torque	Corr. HP	Vol. Eff. %	BSFC
2000	320.8	122.2	83.3	.56
2500	335.9	159.9	88.5	.56
3000	355.3	203.0	91.8	.52
3500	361.2	240.7	92.8	.55
4000	359.5	273.8	93.4	.53
4500	349.5	299.5	91.9	.55
5000	329.0	313.2	90.1	.62
5500	300.5	314.7	86.0	.56
6000	269.3	307.7	80.8	.56

CLOSE CALLS

Component	Clearance (inches)
Main bearing	.0015-.0025
Rod bearing	.002-.0025
Rod side	.009-.017
Crankshaft	.003-.007
Piston pin to rod	.001
Piston pin to piston	.0006-.0008
Piston to bore (cast)	.0015-.0025
Ring end gap: top ring	.014-.016
Ring end gap: second ring	.010-.012

After Bob Lambeck finished with his balance and blueprinting magic, 'Cuda Connection painted the 340 and we bolted it to Edelbrock's Super-Flow dyno. Louis Hammel pulled the handles and pushed the buttons, proving that the A-block could make over 300 horsepower in stock configuration.

PARTS GALORE

Perfect Circle Engine Parts

211-2084	Intake valves (8)
211-2085	Exhaust valves (8)
VS-21	Valve seals (16)
216-5177	Valve locks (16)
229-1654	Camshaft
213-1679	Hydraulic lifters (16)
212-1174	Valvesprings (16)
215-4111	Pushrods (16)
230-3029	Rocker arm shafts (2)
214-2069	Left rocker arms (8)
214-2070	Right rocker arms (8)
9-2103	Matched chain & sprocket set
224-2215 .030	Left pistons (4)
224-2216 .030	Right pistons (4)
40900 .030	Moly ductile ring set
219-3052	Expansion plugs (12)
219-9420	Expansion plug
219-2110	Expansion plugs (12)
601-1018	High volume oil pump
601-1523	Oil pump pickup
228-1482	Water pump
PCB-481P .010	Rod bearings
PMS-540M .010	Main bearings
PSH-325S	Cam bearings
220-1115	Engine mounts (2)

Victor Gaskets

FS-3536	Full gasket set
3536VM	Stainless core head gaskets (2)
95027	Hi-flow exhaust gaskets
VS-39569H	Hi-temp valve cover gaskets

tached sidebar, Bob Lambeck's efforts truly proved themselves when the small-block topped the 314-hp mark.

With a strong 340 and a fully prepared drivetrain, our Swinger is finally coming together for that 2000-mile trek to the Street Machine Nationals. After that, it'll be up to some lucky CC street machiner to cruise it home. Next month we take on the body work and suspension. ℂ

We were lucky enough to find a new service replacement carburetor for the Carter AVS that was originally installed on the 340. According to Carter, there are a few left under the PN 4639S, or Chrysler PN 2946590.

SOURCES

Carter Automotive
Division of ACF Industries
9666 Olive Street Road
St. Louis, MO 63132
314/997-7400

'Cuda Connection
10111 Garden Grove Boulevard
Garden Grove, CA 92641
714/638-2003

Edelbrock Corporation
411 Coral Circle
El Segundo, CA 90245
213/322-7310

Bob Jennings Dyno Tune
9517½ Sepulveda Boulevard
Sepulveda, CA 91343
818/894-3811

Bob Lambeck Enterprises
7510 Gloria Avenue
Van Nuys, CA 91406
818/787-5678

Perfect Circle/Victor
P.O. Box 455
Toledo, OH 43692
419/866-7889

Chrysler Hemi

By Jeff Smith

From Top Fuel to Super Stocks, from the NASCAR circuit to the streets, there are few performance endeavors that the Chrysler Hemi has not dominated at one time or another. Consequently, the elephant motor could justly be crowned the King of the Great American Engines. Looming at the top of the horsepower hierarchy, the

The King of the Chrysler Engine Arsenal

Hemi, as both a race and street runner, is probably the one engine that all enthusiasts hold in reverence.

From its inception, the Hemi was built for one purpose—maximum horsepower. True, the early small-displacement engines were bulky and not the barnstormers that their latter-day brethren were, but they were still the best of their day. The first Hemis found their way into countless rail cars in the early '60s. Even today, with the Hemi out of production for over eight years, it is the basis for current Top Fuel engines, and still the one to beat in

the Super Stock ranks. On the street, it's best to be realistic about your chances when challenging a Mopar powered by the engine with the expansive valve covers and the production 8-barrel induction.

The Hemi legend began with the introduction of the 180 horsepower 331-incher in 1951. Evolving through a number of changes as both a standard and "raised block" version, the old-style Hemi grew to its familiar 354 and 392ci forms that gained prominence in the early fuel racing categories. The Hemis were one of the few engines that could take the pressures of nitro-burning abuse without ending up as cast iron paperweights. By the late Fifties, the early Hemi sported 390 horsepower options under the Chrysler 300C and 300D nameplates—but the original Hemis were dropped as extras at the end of 1958.

Although gone, the Hemi was not forgotten. As the escalating Detroit wars of the mid-'60s demanded larger and more powerful motors, Tom Hoover and the corporate Chrysler clan revived the hemispherical iron. The late-model Hemi, so designated because of its new design and lack of interchangeability with the older Hemis, debuted in 1964 as a "race

only" powerplant that sent the competition scurrying for a new combination to beat the monster motor. The original race Hemi for the NASCAR tracks had cast iron heads and a single 4-barrel aluminum intake manifold. The drag race version Hemi came out the factory doors with an aluminum 8-barrel intake system and an emphatic suggestion that the engine be run only for short 15-second bursts. The drag engines were available with 11:1 or 12.5:1 compression, with only a 10hp difference between the two.

The following year, race Hemis changed little, although the drag race engines were blessed with aluminum heads, magnesium dual-quad intakes and aluminum water pump housings. The NASCAR engines, however, retained their cast iron heads.

The magical year for the Hemi had to be its "detuned" introduction to the street set in 1966. The dreams of Mopar street enthusiasts were realized with the Hemi's availability in the new model year Mopars. Unleashing the Kong motor on the public required certain modifications in the form of a return to cast iron heads, a milder 10.25:1 squeeze factor, a milder solid cam and an inline dual 4-barrel intake with twin Carter carburetors. Even with these changes, Chrysler rated the street Hemi at a conservative 425hp, although the engine's true output was considerably higher.

It was a competitive package, as racers like Jere Stahl, Dick Landy, Sox and Martin and others devastated the large displacement classes and claimed the eliminators as Mopar territory. Even now, Randy Humphrey and Jim Kinnett are again competitive with the Hemi in the Pro Stock ranks.

Except for a slight change to hydraulic lifters in 1970, the street 426 remained relatively unchanged through its eight-year reign as Chrysler's supreme powerplant. Certainly one of the Hemi's more majestic moments was its optional availability in the new Barracudas in 1970 and '71 with the Super Track Pac options that produced 4.10-Dana-geared drivelines that

Magnesium Menagerie

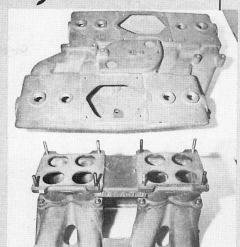

Most late-model Hemi engines came with a dual 4-barrel intake manifold and either Holley or Carter AFB carburetors. The 1964-'65 and '68 race Hemis used Holley carburetors mounted on an A-990 magnesium cross ram intake. The Direct Connection folks recommend this manifold (still available under part number P2536900) for bracket cars and Super Stock A&B classes, along with some type of fresh air hood scoop such as the '69 Road Runner bolt-on scoop P4120199. For an all-out Modified or Pro Stock scoop, use a boundary layer type and be sure to seal the carburetors to the hood. The Direct Connection catalog lists two race manifolds for these applications, the Weiand I.R. and the Weiand tunnel ram, P3690233 and P3690234, respectively.

Years to Ya!

CID	Bore & Stroke	1956	57	58	59	60	61	62	63	64	65	66	67	68	69	70	71
354	(3.94 x 3.63)	▬	▬														
392	(4.00 x 3.90)		▬	▬													
426	(4.25 x 3.75)							Race Hemi		▬	▬			▬			
426	(4.25 x 3.75)							Street Hemi					▬	▬	▬	▬	▬

Ebony Elephant

Return with us now to those thrilling days of yesteryear, when the mastodons of Chrysler's engine arsenal roamed the streets stalking suitable prey. Jim and Ronda Schuett of DeMott, Indiana, have one of these rare specimens encased in a two-car garage. Except for the tall A&A hood scoop that adorns the '66 Coronet 500 body, only the seven small letters and numerals notched in the car's flanks spell out a message that this gunner is more than just another torsion-bar wonder.

Jim made very few changes to the car, and with good reason; the Hemi requires little in the way of improvements. A sprinkling of steel braided lines, a Mopar magnetic impulse ignition with an Accel coil, and Hooker headers help the otherwise inwardly stock Hemi perform to perfection. Cragar five-spokes fit the Hemi's classic stance and are suspended by Direct Connection Super Stock springs along with Monroe shocks on all four corners. A Chrysler 4-speed and Hurst shifter along with a 4.10 rear gear complete the powertrain. You can bet that the streets of DeMott rumble just a little when this street gunner hits town. After all, bad guys wear black, don't they?

Head Tricks

The only thing more impressive than a Hemi-fied street machine is a Hemi with aluminum heads. The original race Hemi that debuted in 1964 came in two versions, one for the roundy-round boys and one for the drag racing clan, with both engines outfitted with cast iron heads. By 1965, however, the quarter-mile race engines were blessed with aluminum heads (bottom) to lighten the engine package; NASCAR Hemis still came with cast iron heads. Subsequent street Hemi and the 1968 race Hemi had cast iron cylinder heads that used a different valve cover than the original aluminum race Hemi heads. Toward the declining years of street Hemi production, the Direct Connection parts list offered a replacement head (top) that had the trick dual spark plug arrangement, but utilized the more common street Hemi valve cover. This head was offered as an alternative to the cast iron street Hemi head.

All production heads (which were single plug only) were built with the spark plug in the "A" position. However, with the advent of the dual-plug heads, Chrysler discovered

had the strength to match the brutal torque of the Hemis.

Those powerful E-bodies marked the beginning of the end for the Hemi's tour of duty, and it was retired at the end of 1971. It fell by the wayside much like the other giant mastodons that were spawned during the horsepower explosion of the flower generation. Even so, many of these giants live on in street

machines and race cars across the country. A multitude of performance pieces are still being manufactured, with a majority of the parts coming from the Direct Connection. Along with this well-thought-out parts list is a strong tech library of racing bulletins that present a straightforward discussion on how to make your Hemi fly.

The knowledgeable Chrysler

the "B" position produced slightly more power.

Both the original A-990 aluminum race heads and the replacement aluminum dual-plug heads are now out of production and difficult to find.

The only current Hemi head still available from Detroit is the Stage II head, which has a number of modifications made strictly for fuel racing in conjunction with the Keith Black aluminum block.

Dialed In

An easy way to determine dome height with the tall Hemi pistons is to fabricate this simple aluminum spacer that will sit over the raised dome of the piston and mount a dial indicator for precise measurement. For Super Stock class race Hemis, the minimum piston height is .767-inch above the deck, while street Hemis use the .536-inch figure.

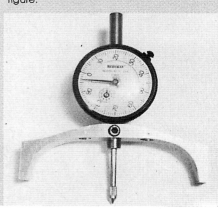

Rocking Tall

Adequate lubrication is a fine point in keeping a Hemi in one piece, whether it's on the street or the strip. One way to avoid the constant chore of replacing rocker arms due to burned pushrod sockets when using the 2-inch valve gear is to drill a secondary lubrication oil hole in the intake rocker arms. The photo illustrates the position of the 3/32-inch oil hole. When marking the position of this hole, be careful not to index the second hole on the rocker arm in line with the primary hole or excessive oil bleed-off will result in reduced oil pressure. Drill the hole all the way through the rocker arm and then fill the outside hole with epoxy. Then remove the outside two threads on the rocker with a mill or grinder to allow the oil to leak down to the pushrod socket.

Some Pro Stock racers even go so far as to use stainless steel rocker shafts with a smaller ¼-inch hole drilled the length of the shaft to replace the stock rocker shaft.

engineers maintain that the trick to setting up a Hemi is really no trick at all. According to the folks who should know, everything that is needed to make the engine strong is designed right into it at the factory. After all, the engine was designed first as a race motor, and then "decreed" to be a street piece.

The Hemi was never intended to be a high production item from the Chrysler option list. Preordained to fill the need for a maximum output motor that also managed to find its way onto the street, the Hemi will always be the one to beat in any kind of competition. If that statement seems somewhat brash, think about it the next time you pull up beside an elephant-powered Mopar. Just be careful not to get stepped on. ◼

Pump Handles

One weak spot in the Hemi oiling system is the oil pump driveshaft. There are two different length shafts that can be used, and if the wrong one is selected, dire consequences result. The stock production shaft is square cut at the drive tang end, and will break off if used in performance circles. The hardened and filleted shaft (center) is also standard length, but will take high rpm abuse. A third shaft (right) is built for the Milodon oil system available for the Hemi and is longer than the other two shafts.

If the longer Milodon shaft is used with the standard oil pump, the longer shaft will break the bottom of the oil pump when tightening down the distributor. On the other hand, if either short length shafts are used in conjunction with the Milodon oil system, the drive tang will not engage the oil pump sufficiently and the drive hex will break off.

The long hex shaft at the bottom of the photo is a simple oil pump primer made by welding a nut on the end of the shaft.

Restricted Area

Most of the tips included in this section have to do with keeping the Hemi together rather than building more horsepower, for the simple reason that the motor is such a brutal performer right out of the box. In this light, many engine builders install bushings to properly align the lifters to the rest of the valvetrain. However, bushing the first two tappet holes on the engine's right side (No. 2 cylinder) also drastically reduces oil to the last three main bearings. A quick sight check down the right side oil galley with the rear plug removed demonstrates the restraint that the bushings present. If you feel the bushings are a necessity, they can be reamed out after installation to allow full oil flow.

Weight Breaks

With the entire country on a physical fitness kick, maybe it's time to put your Hemi on a diet, too. Some racers are resorting to machining the block along its flanks to reduce some of the cast iron "flab" that constitutes extra weight. Caution should be exercised here, though, as too much trimming could get into the water jackets and leave you with a boat anchor instead of a Pro Comp performer.

Along with the obvious advantage of the magnesium intake and aluminum heads is an aluminum water pump (P2536086) that will fit the Hemi and shave a few more precious pounds off the front end.

Elephant Talk

PART NO.	COMPONENT
P3690437	Connecting rod, 396ci 6.960-inch length with ½-inch SPS bolts and nuts
P2806293	Connecting rod, 404ci 7.174-inch length with ½-inch high-strength bolts and nuts
P2531277	Connecting rod, 426ci 7.061-inch length, NASCAR forging with ½-inch high-strength bolts and nuts
P3571046	Piston, right, for 4.255-inch bore
P3571047	Piston, left, for 4.255-inch bore
P2468600	Intake valve, 2.25-inch for iron head
P2531115	Intake valve, 2.25-inch for aluminum head
P3614082	Intake valve, 2.25-inch, use only with 2-inch valve gear
P2468605	Exhaust valve, 1.94-inch for iron head
P2531116	Exhaust valve, 1.94-inch for aluminum head
P3614087	Exhaust valve, 1.94-inch, use only with 2-inch valve gear
P4120071	Valvespring, Street Hemi
P4120072	Valvespring, Purple Stripe, up to .620-inch lift
P4120073	Valvespring, "Battleship"
P4120083	Valvespring, "Chrysler Triple"
P4007851	Rocker arm, intake, adjustable, use only with 2-inch valve gear, 1.7:1 ratio
P3690184	Rocker arm, intake, adjustable standard ratio
P3690185	Rocker arm, exhaust, adjustable standard ratio
P4120265	Rocker shaft support package, standard valve gear
P4120266	Rocker shaft support package, 2-inch valve gear
P4007284	Pushrod kit with assembly tool
P3690814	Camshaft, Racer Brown SSH-44 hydraulic
P4007945	Camshaft, Cam Dynamics CHR-820 hydraulic
P3574029	Camshaft, factory mechanical
P4120263	Roller timing chain and gears, 3-bolt
P4120264	Roller timing chain and gears, 1-bolt
P3690936	Cam, offset bushing package
P2536900	Intake manifold, A-990 magnesium cross ram
P3690233	Intake manifold, Weiand "I.R." magnesium
P3690234	Plenum top for Weiand manifold
P4007177	Long rotor oil pump package
P3571071	Oil pump drive, hardened tip
P3412064	Oil pump drive for Milodon system
P3690875	Oil pump drive for roller cams
P2536086	Water pump housing, aluminum

"B" MOTOR RECIPES

DC's "Shake and Bake" Dyno Packages Bring the 440 to a Boil!

By Marlan Davis

Chrysler's Direct Connection and Arrow Racing's dyno floggers are at it again. This time it's the 440 "RB" big-block, and the folks at Arrow were "microwaving on high" as they re-tested all the factory and aftermarket hardware to come up with dyno-proven, race-winnin' packages. The final product is a revised cookbook of greasy thumb recipes for Mopar chefs that's guaranteed to give them a wedge up on the competition—without driving a wedge through their wallets.

For the tests, Arrow Racing assembled a .030-over mule motor that would survive 88 runs in a horsepower escalation that started at a sedate 300 and kept going all the way to 600-plus. A production block with a late-model *cast-iron* crank was used—just to prove everyday Mopar pieces could take the strain. Lightweight 11:1 Venolia pistons are hooked to the crank by Bill Miller aluminum rods. These rods have an excellent reputation

for durability, and are a bunch cheaper than fully prepared steel con rods. A conventional Speed-Pro plasma-moly ring set ensured continuous good sealing for the long-term dyno flog. Naturally, Arrow meticulously blueprinted and assembled the block (for specs, see sidebar).

A projected e.t. "bracket" is listed for each horsepower combo; "13½-second bracket" means that a B-engined car of the approximate weight listed on the recipe card should be able to turn low 13's in the quarter, assuming an average chassis setup and moderate tuning ability. As engine horsepower increases, we assume that appropriate chassis mods are made to "keep up" with these increases, and that the car gets progressively lighter. More info can be found in DC's "recipe books"—*Engine Speed Secrets* (P4349340) and *Chassis Speed Secrets* (P4349341). Especially useful in predicting e.t.'s is the latter book's Chapter 33, "Bracket Racer Science."

Illustration: Dave Deal

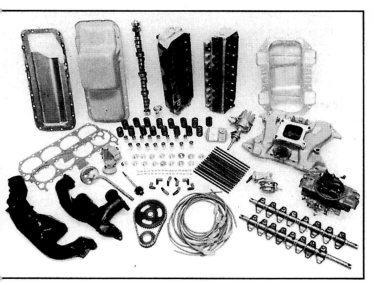

PHASE 1

Every test has to begin somewhere; in this case, a "high torque" 440 was assembled using the previously described short-block and the components listed at left. While not strictly necessary at this stage of modification, high-performance oil and valvetrain components were added to ensure future durability under increasing horsepower levels. Because Chrysler hydraulic rocker arms have no provisions for adjustment to compensate for changes in valvetrain geometry caused by high-lift cam profiles, valve stem height changes, or milled blocks and heads, relatively affordable adjustable pushrods were added to compensate for any possible variations. In many cases, they would not be needed.

Two different hydraulic cams were tested with this basic combination. P4286677, designed for light competition applications, features .455 lift, 112 degrees centerline, 48 degrees overlap, and 272 degrees advertised duration. It offers good mid-range torque and horsepower. The 453-ft.-lb. torque

curve's peak is actually flat between 3000 and 3250 rpm, perfect for a strong-running street or dual-purpose engine. A 350 peak horsepower was observed at 4750 rpm—showing it's easy to make power on this engine without radical mods. Essentially, this package duplicates the old 440 high-performance packages, and should get you into the low 13's. A more streetable 9:1 compression ratio costs about 25 hp, still good enough for low 14's.

DC also offers a very mild "RV" cam, P4286673. This makes more torque and power than the old 383/400 2-barrel cam. With only .410 lift, 248 degrees duration, 20 degrees overlap, and 114-degree centerline, this cam has more area under the torque curve at low speed than the "677" cam, although it is down about 45 hp on top—to 305 at 4000. If you have a heavy passenger car or truck that's driven daily, this is the cam for you, with gobs of power below 3000 rpm. Assuming a 3800-pound car, the cam should deliver 14½-second e.t.'s.

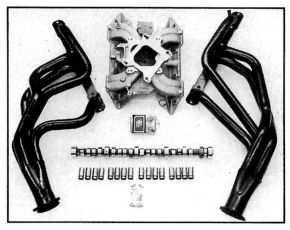

PHASE 2

Your classic hot rod hop-up would be intake, headers, and cam. This is it—bolt on 50 hp. Holley's Street Dominator intake is the best all-round, general-purpose 4-barrel Chrysler intake. Camshaft P4286631 installs on a 110-degree centerline and has an advertised duration of 280 degrees (234 degrees at .050), .474 lift, and 60 degrees overlap. A good set of street headers and axing the mufflers lets the engine breathe freely. More horsepower requires a better ignition system; Chrysler's "Chrome Box" fits the bill for anything besides an all-out race motor.

Net result: While the previous package is all done by 4000, this combo is just starting to come on strong. Yet, thanks to all the cubes and torque, the Phase 2 package is still streetable—in the sense that a 440 Six-Pak or Hemi was streetable (of course, good gas is required). The torque curve is gentle between 3500 to 4500 rpm, with the 478-ft.-lb. peak occurring at 4000 rpm. A 406 peak hp was produced at 5000 rpm—levels the engine will live at all day long. We have here the quintessential dual-purpose bracket setup—race on Sunday, drive it to work on Monday.

PHASE 3

To get into the 11's requires a better cylinder head and a mechanical cam. The resultant engine performance in turn demands a correspondingly improved oiling system and better headers.

Chrysler's Stage V cylinder

head features revised internal port shapes and a combustion chamber engineered to improve airflow characteristics, yet utilizes production valves and valvegear, and is 100-percent interchangeable with all other Mopar cast-iron

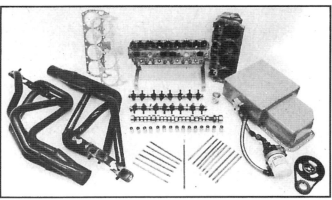

2 Ported DC Stage V heads, P4349661*	10-degree keepers)
16 "Chrysler Triple" valvesprings, P4007536	1 mushroom camshaft and lifter package, P4349270
16 titanium valvespring retainers, P4286775 (for 2-inch spring and	1 Holley R4781 850-cfm double-pumper carburetor, P4349232

*Comes assembled with oversize valve: 2.14-inch intake P4007942, 1.81 exhaust P4120579; P4286612 10-degree keeper; P4120785, 10-degree titanium retainer; and P2806077 valvesprings.

heads (except rare early '60s max wedge heads). The larger Stage V intake port may require some slight grinding for pushrod clearance.

DC carries no purpose-built B-engine full-race oil pan, although Milodon, Hamburger, and Moroso all have excellent units, and the DC race books describe in detail this type of pan's fabrication. Such a pan, plus the addition of the special Milodon external oil pump assembly with swivel pickup, ensures reliable oiling under all conceiveable drag racing conditions.

Because the B engine's hydraulic rockers aren't adjustable, mechanical rockers are required when installing a solid lifter cam. (These special rockers can also be used to provide an adjustable hydraulic camshaft valvetrain, but adjustable hydraulic pushrods are cheaper.) Since the mechanical cams use a stronger 3-bolt upper cam sprocket in place of the 1-bolt attaching system used with the current DC hydraulic grinds, Arrow installed a new 3-bolt upper sprocket.

Two different mechanical

The .015-thick spacers contained in kit P3690896 are used to improve adjustable rocker arm-to-valve tip alignment (arrow A). With the oversize valves, 2-inch valvesprings, and special retainers, underside of rocker may need relieving (arrow B).

cams can be used for this phase. If the car is to be street-driven at all (and you'll need steep gears, a big converter, and manual brakes), cam P4120659 is recommended. It specs out at .528 lift, 284 degrees duration, 60 degrees overlap, and installs on 112-degree centerline. Result: 446 hp at 5500 rpm, with the torque peak occurring at 4250 rpm, where 488 ft.-lbs. is produced. Both torque and horsepower curves were moved up the scale about 500 rpm, but they're still not peaky. The relatively flat hp curve makes this grind a great all-round cam with either an auto or manual trans. All in all, we're talking about an 11½-second car here.

To knock off another half second requires about another 50 hp. The extra 50 hp crosses the dividing line from marginally streetable to completely unstreetable. Nevertheless, the added 50 hp can be found with merely a cam swap. P4349268 has .620 lift, 324 degrees duration, 112 degrees overlap, and 106 degrees centerline. This combo produces 486 hp at 6000 rpm, with 478 ft.-lbs. at 4750. But look where the torque and power come on. Previously, each step up resulted in a package that more or less outperformed the previous combination over most of the mid to high-rpm range, with some sacrifice at the low end. This one's different. Peak torque is now actually 10 ft.-lbs. less than the previous milder cam, and it occurs 500 rpm higher. Below 4000, there's no power at all. The horsepower level only exceeds the previous grind's performance past 4250 rpm. See what we mean by "race only?"

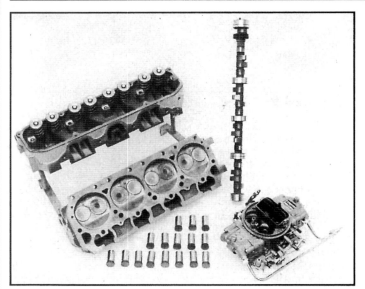

PHASE 4

When you get to this performance level, we're talking all-out race. "All-out" usually means no-nonsense components—roller cams, maybe even exotic aluminum heads. But one of the 440's virtues is its performance per dollar when compared to other big-blocks. Roller cams and exotic heads add thousands of dollars to an engine's cost, while decreasing reliability. Is there a cheaper way?

Yes. Recently DC introduced a ported and polished Stage V head assembly that comes fully assembled with oversize valves, titanium retainers, and race valvesprings. If you shop around, you can buy them for about $1000 apiece. Unfortunately, the as-installed valvesprings coil-bind with the Phase 4 cam, so they were replaced by the famous "Chrysler Triple" valvespring, along with a special titanium retainer designed to accommodate these 2-inch installed height valvesprings on standard-length valves.

As for the cam itself, DC

now carries several "mushroom tappet" designs—the poor man's roller. The largest cam (P4349270) has .690 lift, 328-degree duration, and 107-degree centerline. It offers performance quite close to a roller—but without a roller's expense and complexity. It will do the job in any application besides all-out "pro" racing.

Several block mods are required to use the ported heads and mushroom cam. Bore notches are usually required to clear the 1.81-inch oversize exhaust valve, while the bottom of the lifter bores should be spot-faced to clear the mushroom tappets' larger "foot." A larger 850 carb is also recommended, since at this performance level it's worth 10 hp over the 750.

Looking at the dyno results, it's apparent we're totally in high-rpm territory now. All power is on the top end. Yet the cam has a broad flat torque and horsepower curve. It works well not just at the drags, but also in oval track competition.

PHASE 5
9.90-SECOND BRACKET—603 HP
3000-pound car, all-out suspension and driveline, 14-inch slicks

- 1 Weiand tunnel ram, P4120790
- 2 Holley R4779 750-cfm double-pumper carbs, P4007900*
- 1 set fabricated race headers—2⅛x36 primaries into 3½x15 collector
- 1 Super "Gold box"
- 1 ignition module, P4120600
- 1 Accel race coil, P3690560
- 1 dual-carb linkage kit, P4349467
- 1 Holley GPH-110 high-volume electric fuel pump and regulator kit, P4120227

* Re-jet carbs to No. 70 on all four corners

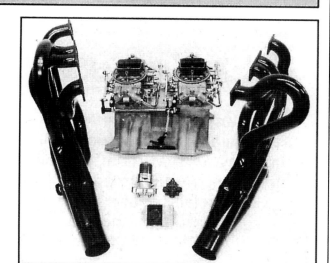

TEST RESULTS

[Graph showing HORSEPOWER vs ENGINE RPM and TORQUE vs ENGINE RPM]

Horsepower curves:
- Phase 5 — 603
- Phase 4 — 560
- Phase 3A — 486
- Phase 3 — 446
- Phase 2 — 406
- Phase 1A — 350
- Phase 1 — 305

Torque curves labeled: Phase 5, Phase 2, Phase 3A, Phase 3, Phase 1A, Phase 4, Phase 1

Horsepower axis: 250–650
Torque axis: 350–550
Engine RPM axis: 3000–7000

PHASE 5

When the tests began, DC's goal was to achieve 600 hp with a non-roller cam. At that power level, a 3000-pound car with automatic trans can run NHRA 9.90 Super Gas classes. By adding a dual-carb tunnel ram setup, larger headers, and the all-out "Gold Box" electronic ignition module, DC engineers achieved their goals, ending up with 603 hp at 6500 rpm, with a torque peak of 527 ft.-lbs. at 5500 rpm—and this on a dyno mule with 87 runs already under its belt.

Other than re-jetting the carbs for the final test, it's important to remember that all these parts were run as-is out of the box. The intakes weren't even port-matched. With a little fine-tuning, there's bound to be more horsepower on tap waiting to be unleashed.

These packages work because they are a coordinated combination—each should be installed in its entirety, or the actual performance gains will be disappointing. The 413 or 426-based raised deck B motors will have similar results to the 440 used here for the development program. If you're fortunate to have an early-'60s max wedge engine, you're a leg up to start. The smaller low deck 383/400 engines may require a different package to achieve equal performance gains.

Chrysler 440's are a dime a dozen in the salvage yards. These tests proved they don't need trick one-of-a-kind pieces to get the job done. And thanks to DC's latest cookbook, there's no guesswork involved. Follow the recipe and bake—then your 440 will thunder and shake. **HR**

DYNO TEST ENGINE SPECIFICATIONS
ENGINE BUILDER: Arrow Racing Engines, 224 South St., Rochester, MI 48063, (313) 652-0604.
ENGINE TYPE: Chrysler OHV V8, Type RB, 446 cubic inches (440+.030), 4.35 bore x 3.75 stroke, 11:1 compression ratio.

MAJOR INTERNAL COMPONENTS
Parts not listed below: see descriptions for each phase.

BLOCK	Seasoned stock production, bored .030 and honed with torque plates using Sunnen 625 stone
CRANK	Production cast iron, internally balanced with Mallory heavy metal
BALANCER	DC P3830183 thin forged damper
PISTONS	Venolia lightweight forged flat-tops
RINGS	Speed-Pro R-9224+.035: 5/64 plasma-moly 1st, 5/64 plain cast-iron 2nd, 3/16 standard tension oil
MAIN BEARINGS	DC P4286976 babbit
ROD BEARINGS	DC P2836184 babbit
CONNECTING RODS	DC P4349221 Bill Miller aluminum
OFFSET CAM BUSHING SET	P3690936 (for centerlining cam)

BUILD SHEET
All dimensions in inches or fractions thereof, unless otherwise indicated. Data not listed below: refer to stock service manual or DC race manual.

ROD/MAIN BEARINGS	.0025
CRANK END-PLAY	.006
CON ROD CENTER-TO-CENTER	6.768
CON ROD SIDE CLEARANCE	.025 (aluminum rods)
PISTON-TO-WALL	.006
RING GAP	
1st	.012
2nd	.010
PISTON DECK HEIGHT	+.022 (above deck)
PISTON-TO-VALVE CLEARANCE (0 lash)	
Automatic trans	.090 intake, .100 exhaust
Manual trans	.100 intake, .110 exhaust
PISTON-TO-HEAD	
(with gasket)	.075 (aluminum rods)
CYLINDER HEAD VOLUME	
Except Stage V head	85.5cc
With Stage V head	89cc
GASKET COMPRESSED THICKNESS	
Composition P4349559	.040[1]
Steel shim P4286754	.017[2]
VALVESPRING INTALLED HEIGHT	
All except P4007536	1.86
P4007536	2.00
VALVE LASH	
All hydraulic cams	0
P4120659 & P4349268 mechanical	.028 intake, .032 exhaust
P4349270 mushroom	.024 intake, .028 exhaust
ROD BOLT TORQUE	
(P4349221 rod)	85 ft.-lbs.
SPARK PLUG GAP	
(Champion J63Y)	.035
TIMING	38° total

[1] Used to achieve 11:1 CR with 85.5cc chamber head.
[2] Used to achieve 11:1 CR with 89cc chamber head.

Few men become legends in their own time. Keith Black is one of the few. The best-known engine builder of Top Fuel and Funny Car drag racing engines, Keith and his South Gate, California, facility are also the main suppliers of drag racing engine parts in the world. If you want to go Top Fuel racing, you talk to Keith.

In recent years, however, Keith has been diversifying his company, hiring Dan Borre to head his street engine operation. And to kick off his new venture, Keith gave us the lowdown on how to build the best small-block Mopar for the money.

"Laying it on the line" right from the start, Keith felt that there were certain points which all street engine builders should note. First, for $1000 you won't have the ultimate engine. More than likely you will have an engine that's a little stronger than stock. In addition, any time you alter an engine from stock there is a "trade-off."

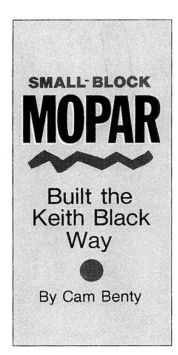

SMALL-BLOCK MOPAR

Built the Keith Black Way

By Cam Benty

According to Keith, "the factory did an excellent job researching and building an engine which had good durability and power. As soon as you alter it from stock, you run the risk of lessening the

durability, if only because the engine is running at a higher rpm for longer periods of time. If you want durability, then go with a stock engine rebuild."

Keith is a big fan of the cubic inch. The bigger the engine, the more horsepower potential. Obviously the 360cid engine is his favorite, with the 340 close behind because of its forged crankshaft (Keith feels the 360cid cast crankshaft is up to most street applications, especially when engine rpm is kept below 6000 rpm). For the engine builder with more than a $3000 budget, Keith suggests the use of his stroker kit, which nets close to 391cid from the small-block. But for our purposes, we decided to stay with the original displacement.

All of the parts listed here will fit on either the 360 or 340cid engines. Most of the parts are available through the Direct Connection catalog or through a dealer such as Keith. As an added benefit, Keith offers a 10-percent

Keith Black is the most renowned drag racing engine builder in the world. In Top Fuel and Funny Car competition, his engine is the standard whereby all others are judged.

bulk discount on large orders of Keith Black and Direct Connection parts. All parts listed are available through Keith Black at the quoted prices; however, some prices are subject to change.

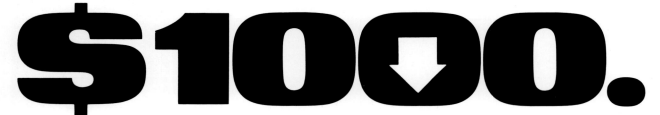

$1000.

As noted in the introduction, for $1000 you're not ready for Super Stock competition, but for basically rebuilding your stock engine. Since you are using the original pistons, crankshaft, cylinder heads, and connecting rods, the performance modifications are oriented more toward fine-tuning. Of particular note is the

camshaft selection, which is a hydraulic unit with gross lifts of 0.420-inch intake, 0.442-inch exhaust, durations at 0.050-inch lift of 208 degrees intake and 218 degrees exhaust. Lobe center is 112 degrees. In addition, new valvesprings and lifters are a must with any camshaft change.

Keith recommends dual-plane manifolds for street applications. The Weiand 2P is his choice for engines running low rpm and the Edelbrock LD340 for higher rpm applications. With either manifold and any of the engines in this article, Keith feels the Holley 650 cfm carburetor is more than up to the task. "Even with the $3000 engine, a street motor rarely requires more than 650

Keith offers his own brand of camshafts with streetable lifts and durations. The camshaft for the $1000 and $2000 ▷

▷ engines are both hydraulic. Holley 650 cfm carb will flow ample air and fuel for all three of the engines' requirements.

● MOPAR ● MOPAR ● MOPAR ● MOPAR ● MOPAR ● MOPAR ● MOPAR ● MOPAR ●

cfm of air," said Keith.

Although not noted as part of the engine build-up, headers are a major tuning device and selection is very dependent on the application. Good results can be had with small 1⅝-inch primary tube headers. The weight of the car and the type of engine being built are deciding factors here. In the ignition department, Keith likes the stock Chrysler electronic system that has been around since the late 1960s. His only oiling system modification is the installation of a high-pressure Direct Connection spring at a cost of $2.

Keith recommends use of dual-plane manifold with all Mopar street engines. Edelbrock LD340 should be used with higher-rpm engines, while Weiand 2P is the manifold of choice for lower-rpm operation.

The stock Chrysler electronic ignition is Keith Black's choice for all engines, as it has the power to ignite almost any engine, especially those with 9.5:1 compression or less.

$1000 ENGINE

All prices quoted according to Keith Black Racing Engines.

All Direct Connection parts listed at Racer Net price. Keith Black will discount for bulk purchases.

PART	PRICE
KB Camshaft PN 4120040; Lifters and Lube	$172
KB Valvesprings PN 450839	63
Weiand 2P PN 8007 or Edelbrock PN LD340	145
Holley 650 cfm PN 4777	195
KB Fuel Pump PN PPF-12	85
Speed-Pro rings (moly face)	115
KB engine blueprinting	125
K&N air filter	35
DC oil spring PN P3690944	2
DC Timing gear and chain PN P3690866	50
TOTAL	$987

$2000.

For $2000, you can build quite a bit more engine. The biggest advantage is the addition of the W-2 cylinder heads which flow much better than the standard "A" engine heads and can be ported much larger than the original heads, due to repositioned intake ports and pushrod holes. Unfortunately, the W-2 cylinder heads require many additional parts, such as special valves, rocker arms, rocker stands, and rocker shafts along with a different head bolt set. But according to Keith, the performance value of the heads is worth sacrificing the air cleaner filter and fuel pump, the stock units probably being up to the task at hand. If you can afford the parts, buy them.

W-2 cylinder heads require completely different rocker arm gear than stock A-engine; however, they offer far greater performance potential.

AR ● MOPAR ● MOPAR ● MOPAR ● MOPAR ● MOPAR ● MOPAR ● MOPAR ● MO

With the added performance from the cylinder heads, Keith recommends the second-level KB camshaft and lifter set, which sells for the same price as the $1000 engine camshaft making the swap a ''no charge'' modification. The level two camshaft is also hydraulic and has gross lifts of 0.441-inch for the intake and exhaust with durations (again measured at 0.050-inch lift) of 218 degrees intake and 228 degrees exhaust. Lobe center is 114 degrees.

The most subtle change here is the machine work necessary to mate the intake manifold to the W-2 cylinder heads. As neither Weiand nor Edelbrock make a dual-plane manifold for the heads, Keith Black will adapt your manifold by welding up the original bolt holes and redrilling for the new heads for only $40.

Exhaust port comparison between W-2 heads (above) and stock A engine heads (below) is most graphic example of performance difference.

$2000 ENGINE

Includes same components as $1000 engine with following additions and modifications.

With new cylinder heads, manifold must be modified; note original bolt holes on Weiand intake.

PART	PRICE
Second-level Keith Black camshaft PN P4120180	$0
DC W-2 cylinder heads, bare PN P3870812	600
DC Head bolt kit PN P4007712	6
DC W-2 Long stem 2.08-inch intake valves, PN P4007941	90
DC W-2 Long stem 1.60-inch exhaust valves PN P3614589	90
DC aluminum retainers PN P4286573	54
DC valve stem single groove locks PN P4120618	22
DC Econo W-2 Rocker Arm PN P3870825, PN P3870826, PN P3870827	201
DC Rocker arm shaft support set PN P4120102	90
DC W-2 rocker shafts PN P3677086	75
Machine work necessary to make dual-plane manifold compatible with W-2 system	40
TOTAL	1268
Less 10 percent bulk discount on DC and Keith Black parts	<126>
LESS:	
K&N air filter	<35>
KB fuel pump	<85>
PLUS:	
Cost of first engine	987
NEW TOTAL	$2009

$3000.

Now it's time for serious engine building. If you set your budget at $3000, (auxiliary costs for machine work and rebuild parts such as gaskets, bearings, belts, hoses, and headers could bring your total cost closer to $4500) the engine you build will probably have more horsepower than required for the street. At this level, the engine is still docile enough for daily driving, but comes close to the limit, necessitating constant attention and service for optimum performance. This is *not* a 100,000-mile engine, so don't be misled.

The biggest performance additions here are installation of stronger forged aluminum pistons and tougher connecting rods. The new W-2 cylinder heads will net their most value when you open them up for better flow. Keith Black offers porting from $150 to $500, dependent on your budget. How fast you want to go is totally controlled by your budget.

With this engine, a high-performance fuel pump is essential along with a step-up in the ignition department to an Allison coil (Allison plug wires could also be of help here if you can spring for the few extra bucks). But overall, the engine requires only your attention to be a world-beater. ⊛

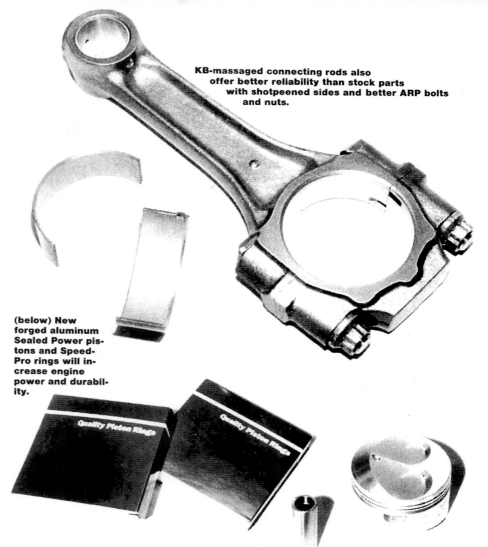

KB-massaged connecting rods also offer better reliability than stock parts with shotpeened sides and better ARP bolts and nuts.

(below) New forged aluminum Sealed Power pistons and Speed-Pro rings will increase engine power and durability.

Porting and cylinder head work is where the biggest performance gains are to be had. How fast you want to go is totally dependent on how much money you have.

$3000 ENGINE

Includes same modifications and parts as $2000 engine with following changes:

PART	PRICE
KB fuel pump	$85
Allison ignition coil	50
KB forged aluminum pistons, wrist pins & locks	450

KB connecting rods (reconditioned, shotpeened, and Magnafluxed) with bolts and nuts	250
Port and polish cylinder heads (medium port job)	300
TOTAL	1135

LESS: KB discount of 10 percent for bulk purchase	<113>
PLUS: $2000 engine	2009
NEW TOTAL	$3031

SOURCES

Keith Black Performance, Inc.
Dept. CC
2622 Risa Drive
Glendale, CA 91208
818/244-2031

Allison Automotive Company
Dept. CC
720 East Cypress Avenue
Monrovia, CA 91016
818/303-3621

Direct Connection
Dept. CC
P.O. Box 1718
Detroit, MI 48288
313/497-1220

Edelbrock Corporation
Dept. CC
411 Coral Circle
El Segundo, CA 90245
213/322-7310

Holley Parts Division
Dept. CC
11955 East Nine Mile Road
Warren, MI 48089
313/497-4000

K & N Engineering Inc.
Dept. CC
P.O. Box 1329
Riverside, CA 92502
714/684-9762

Speed-Pro/Sealed Power
Dept. CC
100 Terrace Plaza
Muskegon, MI 49443
616/724-5011

Weiand Automotive Industries
Dept. CC
P.O. Box 65977
Los Angeles, CA 90065
213/225-4138

MOPAR ● MOPAR ● MOPAR ● MOPAR ● MOPAR ● MOPAR ● MOPAR ● MOPAR ● MOPAR

Building A Better
MOUSE TRAP

The Direct Connection 13-Second Mopar Plan
By Todd Howard

Direct Connection contracted Arrow Racing Engines, an automotive specialty machine shop catering to Mopars, to build and test this 360 A-engine. The objective was to evaluate the parts combination proven to power a street Mopar down the quarter in the 13's.

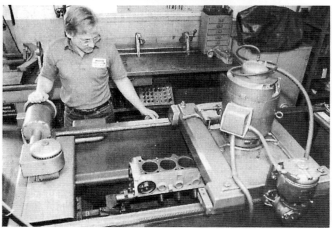

With this in mind, Howard Comstock of Arrow set out to build a competitive, but reliable, street 360. Machining included surfacing the block and heads with a special stone grinder which imparts a smoother surface than the normal cutters.

For many years, the Mopar 340 has been the mainstay of Chrysler's small-block performance heritage. Out of production since 1973, a rebuildable 340 is getting hard to come by. If you're lucky enough to find one, expect to pay a healthy sum for your prize. A budget alternative for the Pentastar A-engine builder is the overlooked 360. Also an A-engine, the 360 is more than capable of setting a trap for an unsuspecting mouse motor.

Besides the cost differential, the prospective 360 builder could benefit in other areas as well. First, since the 360 is a newer model engine used in passenger cars from '71-80, they're more plentiful in the salvage yards. The 360 is also a full 20 cubic inches larger, utilizing a .270-inch longer stroke for more torque potential. Then, with the exception of the crankshaft and pistons, most parts from the 340 will interchange with a 360 giving you a wealth of performance goodies to choose from.

To prove the capabilities of the 360, Direct Connection, Chrysler's performance parts division, recently built a fully streetable 360 using hi-po factory components. Arrow Racing Engines was selected to machine, assemble, and test the Street 360 combination outlined in the Direct Connection catalog. The goal was to build an engine capable of powering most any Chrysler musclecar down the quarter mile with consistent 13-second results.

Arrow Racing's Howard Comstock took on the project and began by prepping the two-bolt main block. While a 360 race block is available from DC, Howard chose to keep costs down and work with a production block since the engine would never see the high side of 6000 rpm.

The first step was to bore, hone, and deck the block. An alignment check of the main bearing saddles indicated that align honing would not be necessary. The block's interior was also deburred, followed by bottom tapping all threaded holes and chamfering the head bolt holes. Howard believes in a completely

MOUSE TRAP

CHART #1

━━━━ Stock 318cid w/2-barrel carb
(240-degree duration, .400-inch lift)

- - - DC P4286667 cam
(248-degree duration, .410-inch lift)

········· Stock 360cid w/2-barrel carb
(252-degree duration, .410-inch lift)

━━━━ DC P4286669 cam
(260-degree duration, .430-inch lift)

CHART #2

━━━━ Stock 340cid w/4-barrel carb
(268 / 276-degree duration,
.445-inch lift)

- - - DC P4286671 cam
(272-degree duration, .455-inch lift)

········· DC P4286630 cam
(284-degree duration, .471-inch lift)

━━━━ DC P4120233 cam
(292-degree duration, .508-inch lift)

sterilized block for optimum piston ring sealing, which means a thorough lacquer thinner cleaning, then progressing to a soap and water wash, followed by a rinse and forced air drying. As a final precaution, Howard wipes the cylinder bores with a lacquer thinner-soaked paper towel to remove the rust film left by the water bath.

Since this engine was intended as a relatively mild street motor, the 360's cast nodular iron crankshaft is entirely adequate. No special operations are required other than a light deburring and a journal polishing. After a crank has been reground to size, you can polish the journals using 600-grit emery cloth and a string of rawhide. The rawhide is used to work the paper across the journal in a uniform fashion.

Direct Connection magnafluxed and shot peened connecting rods were selected to replace the stock arms, since long-term durability was a prime concern. These rods come equipped with high-strength bolts and nuts. It should be noted that the nuts supplied must be installed with the machined side facing the rod for proper torque indication.

Any engine that is intended for the street must take into account gasoline quality, which translates to moderation in compression. For this 360 engine, Howard used TRW forged economy-type pistons rated at 9.5:1 compression. Other than deburring the skirt edges, notching for more valve clearance (per the DC engine manual specifications), and fitting moly-coated rings, the pistons remained stock.

Howard assembled the short-block at this time as outlined by the factory assembly manual. A few important considerations were taken which are worth discussing. First, the rod and main bearings were cleaned with lacquer thinner to remove any grit that may have accumulated. The film on the bearing's inner surface was not removed, however, since this provides important break-in lubrication. When fitting the A-engine's rear main cap, Howard also applies a light coat of liquid teflon pipe sealant under the cap to prevent oil leaks from the block-to-cap parting surfaces. Another tip is to leave the front timing cover loose until the damper is in place. This centers the front seal on the harmonic balancer, since the timing cover isn't equipped with locating pins. And, as with any en-

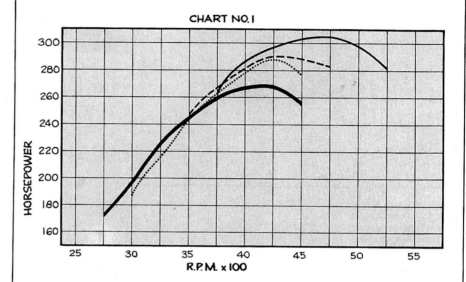

CHART NO. I

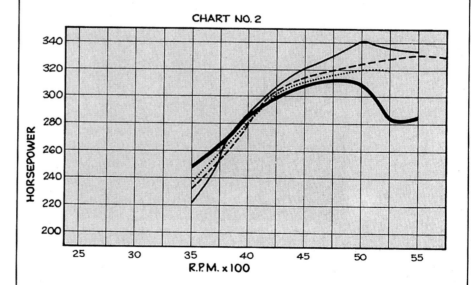

CHART NO. 2

CAM-PARISONS

The main reason for building this streetable 360 was to allow the Direct Connection people to evaluate their lineup of camshafts. While the old Chrysler "Purple Shaft" cams have been around over 19 years, they have not stayed the same. In keeping with this policy of updating their lineup, DC used this street 360 as a mule to test a series of five Purple Shaft cams and three popular factory grinds. As you can see from Graph Number 2, the cam recommended for the 13-second engine package (which also happens to be DC's largest hydraulic cam) made the most horsepower and torque without sacrificing too much street driveability. While 341 hp is respectable, Arrow Racing felt that the engine was limited by the induction system and went on to install a Holley single plane manifold (PN 4007664) and vacuum secondary, 750 cfm Holley carb (PN 4349228) and retested the last two cams. Their results showed an astonishing 349 hp/5500 rpm for the .471-inch lift cam and 374 hp/6250 rpm on the .508 incher.

MOUSE TRAP

gine build-up, both the front and rear main seals should be coated with Lubri-plate to reduce friction upon initial startup.

Howard chose to keep the stock 360 heads, but modified them for increased flow. The factory 1.88-inch intake valves were replaced with larger DC 2.02-inch pieces, while the DC 1.60-inch exhaust valves used measure the same as stock. A standard 45-degree seat angle was cut on the valves, followed by a backcut of 32 degrees, which Arrow has found is worth a 10 percent increase in low- and mid-lift flow. The valve seats were also touched-up with three angles on the intake and four on the exhaust. Both top angles measure 25 degrees, with a standard sealing angle of 45 degrees, and a back cut of 65 degrees. In addition to these angles, the exhaust ports in most A-engine heads respond well to a 75-degree throat angle to improve flow in the bowl area.

With the valves in place, Tim Connolly, Arrow Racing's head porter, performed a mild port and polish job, which included matching the intake and exhaust ports, as well as cleaning up the short-side radius of both runners. Since the ports are relatively small to begin with, Tim usually breaks through to the valve cover bolt holes. To prevent a potential vacuum leak, Howard uses valve cover studs coated with Loctite 271 to seal the holes.

Final engine assembly included fitting the heads and an Edelbrock Performer intake / Thermoquad 800 cfm carb induction system, along with DC's electronically triggered distributor and solid core plug wires. Timing is set at 35 degrees total ignition advance for pump gasoline.

On Arrow's Superflow dyno, Howard found the 360 to produce 341 hp at 5000 rpm and 382 lbs./ft. torque at 4250 rpm. Calculating that out on a dragstrip dream wheel, which assumes optimum traction and gearing, a car weighing as much as 4000 pounds could still run in the 13's with this combination. Lose a few pounds and you have the makings of a serious street contender.

(1)

(2)

(3)

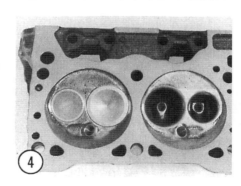

(4)

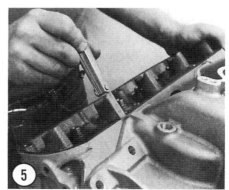

(5)

1. Howard warns that the block should be examined after hot tanking to assure that it contains all of the necessary oil galley plugs. Without this plug, located beneath the rear main cap (arrow), your oil will bypass the filter. Other missing plugs can inflict more disastrous results.

2. Direct Connection's 233 "Purple Shaft" cam plays an important role in obtaining the horsepower for 13-second passes. Arrow also incorporated the DC roller timing chain set, hydraulic lifter package, adjustable pushrods, HD left and right rocker arms, and hi-po valvesprings.

3. For a durable bottom end, Arrow Racing used TRW forged pistons, DC heavy-duty shot-peened rods, and the stock nodular iron crankshaft.

4. The heads used are typical late-model 360 castings which were mildly ported and fitted with larger 2.02-inch intake valves and stock diameter 1.60-inch exhaust valves.

5. If either the block deck or head surfaces are milled, the engine should be mocked up to assure correct intake manifold-to-head alignment. Use a feeler gauge to measure the top-to-bottom and end-to-end variance using only the block rail gaskets located fore and aft. A difference of .005- to .006-inch should be considered the maximum allowable.

6. After final assembly, the engine was tested on Arrow's Superflow dyno. Cylinder pressure leakage was found to be a mere 4 percent, which only dropped to 6 percent after 75 pulls on the dyno. With the biggest DC cam tested, the 360 produced a respectable 341 hp/5000 rpm.

(6)

MUSCLECAR REPORT

Spinning Tires And Yarns About The Musclecar Habit

DIRECT CONNECTION ENGINE PARTS

P3690641	H.D. connecting rods (8)
P3690638	Rod bearings (8)
P4286948	Main bearings-'74-80 360 std.
P4286589	High-volume oil pump
P3690715	Intermediate shaft
P4120233	Hydraulic camshaft
P3614321	Hydraulic lifter (16)
P4120249	H.D. valvesprings (16)
P4286500	Camshaft key
P4286808	Adjustable hydraulic pushrods (16)
P3690710	H.D. rocker arms-right (8)
P3690711	H.D. rocker arms-left (8)
P3690230	2.02-inch intake valves (8)
P3690231	1.60-inch exhaust valves (8)
P4120492	Viton valve seal package
P4120262	Roller timing chain and sprockets
P4286531	Edelbrock Performer intake manifold
P4120889	High-output ignition coil
P4007667	Plug wire separators
P4120716	Solid core spark plug wire set
P3690426	Electronic ignition kit
P4007040	High-output mechanical fuel pump
P4286575	Chrome air cleaner
P4120608	Chrome valve cover set
P4120272	Cover accessory package
P4120446	Chrome breather cap
P4120993	Damper timing tape

OTHER ENGINE ACCESSORIES

N9Y	Champion spark plugs (.040 gap)
9811	800cfm Thermoquad carburetor
	1⅝-inch headers

RECOMMENDED CHASSIS MODIFICATIONS

4.10:1 ring & pinion gears
Sure Grip differential
8x26-inch drag slicks
Shift improver kit-auto trans
Zero toe-in alignment
Tire pressure:
Front 36 psi
Rear 8-16 psi
Adjustable pinion snubber
3800-pound maximum chassis weight

SOURCES

Arrow Racing Engines
Dept. CC
224 South Street
Rochester, MI 48063
313/652-0604

Direct Connection
Chrysler Corporation
Dept. CC
P.O. Box 1718
CIMS 423-13-06
Detroit, MI 48288

HEMI ENGINE PRODUCTION FIGURES

	With 426-8 Total (%)	With 426-8 Auto Trans. Total (%)	With 426-8 4/Spd Trans Total (%)
1965			
DODGE			
Charger	99 (98.0)	93 (92.1)	0 (0.0)
PLYMOUTH			
Belvedere I—2-door	63 (2.1)	60 (1.9)	0
Super Stock	101 (100)	101 (100)	0
1966			
DODGE			
Charger	468 (1.5)	218 (0.7)	250 (0.8)
Coronet—2 door	34 (4.0)	23 (2.7)	11 (1.3)
Coronet—4 door	2 (0.1)	2 (0.1)	
Coronet Deluxe—2 door	49 (1.9)	18 (0.7)	31 (1.2)
Coronet 440—2 door HT	288 (0.9)	96 (0.3)	160 (0.5)
Coronet 400—convertible	6 (0.2)	3 (0.1)	0
Coronet 500—2-door HT	340 (1.0)	136 (0.4)	204 (0.6)
Coronet 500—convertible	21 (0.7)	9 (0.3)	12 (0.4)
PLYMOUTH			
Belvedere I—2-door sedan	136 (6.9)	57 (2.9)	
Belvedere II—2-door HT	531 (1.9)	251 (0.9)	22 (1.1)
Belvedere II—convertible	10 (0.5)	6 (0.3)	280 (1.0)
Satellite—2 door HT	817 (2.6)	314 (1.0)	4 (0.2)
Satellite—convertible	27 (1.1)	**	503 (1.6) **
1967			
DODGE			
Coronet R/T—2 door sedan	59 (2.4)	27 (1.1)	32 (1.3)
Coronet R/T—convertible	2 (1.0)	1 (0.5)	1 (0.5)
Charger—2-door HT	27 (0.7)	8 (0.2)	19 (0.5)
PLYMOUTH			
Belvedere GTX—2-door HT	108 (4.3)	63 (2.5)	45 (1.8)
Belvedere GTX—convertible	17 (7.1)	10 (4.2)	7 (2.9)
1968			
DODGE			
Coronet Sport—2-door HT	125 (1.6)	94 (1.2)	31 (0.4)
Coronet R/T—2-door HT	220 (2.2)	120 (1.2)	100 (1.0)
Coronet R/T—convertible	9 (1.7)	8 (1.5)	1 (0.2)
Charger R/T	475 (2.7)	264 (1.5)	211 (1.2)
PLYMOUTH			
Road Runner—2-door coupe	840 (2.9)	377 (1.3)	464 (1.6)
Road Runner—2-door HT	171 (1.1)	62 (0.4)	109 (0.7)
GTX—2-door HT	410 (2.4)	**	**
GTX—convertible	36 (3.9)	**	**
1969			
DODGE			
Coronet Super Bee—2-dr HT	92 (1.2)	54 (0.7)	38 (0.5)
Coronet Super Bee—2-dr Cpe	166 (0.9)	74 (0.4)	92 (0.5)
Coronet R/T—2-door HT	97 (1.5)	39 (0.6)	58 (0.9)
Coronet R/T—convertible	10 (2.3)	6 (1.4)	4 (0.9)
Charger R/T	432 (2.3)	225 (1.2)	207 (1.1)
PLYMOUTH			
Road Runner—2-door coupe	356 (1.1)	162 (0.5)	194 (0.6)
Road Runner—2-door HT	422 (0.9)	188 (0.4)	234 (0.5)
Road Runner—convertible	10 (0.5)	6 (0.3)	4 (0.2)
GTX—2-door HT	198 (1.4)	98 (0.7)	98 (0.7)
GTX—convertible	11 (2.0)	6 (1.1)	5 (0.9)
1970			
DODGE			
Coronet Super Bee—2-door coupe	4 (0.1)	0	4 (0.1)
Coronet Super Bee—2-door HT	32 (0.3)	11 (0.1)	21 (0.2)
Coronet R/T—2-door HT	13 (0.6)	9 (0.4)	4 (0.2)
Coronet R/T convertible	1 (0.4)	0	1 (0.4)
Charger R/T	112 (1.2)	56 (0.6)	56 (0.6)
Challenger R/T—HT	287 (2.1)	150 (1.1)	137 (1.0)
Challenger R/T convertible	9 (0.9)	4 (0.4)	5 (0.6)
Challenger R/T—HT (SE)	60 (1.6)	37 (1.0)	23 (0.6)
PLYMOUTH			
Road Runner—coupe	74 (0.5)	30 (0.2)	44 (0.3)
Road Runner—HT	210 (0.9)	93 (0.4)	117 (0.5)
Road Runner—convertible	3 (0.5)	2 (0.3)	1 (0.2)
GTX—2-door HT	72 (1.0)	29 (0.4)	43 (0.6)
'Cuda—2-door HT	652 (3.9)	368 (2.2)	284 (1.7)
'Cuda—convertible	14 (2.6)	9 (1.6)	5 (0.9)
1971			
DODGE			
Charger Super Bee	22 (0.5)	13 (0.2)	0
Charger R/T	63 (2.3)	33 (1.2)	0
Challenger R/T 2-door HT	71 (1.8)	12 (0.3)	59 (1.5)
PLYMOUTH			
Road Runner	55 (0.4)	27 (0.2)	0
GTX	30 (1.1)	0	11 (0.4)
'Cuda—2-door HT	108 (2.0)	38 (0.7)	60 (1.1)
'Cuda—convertible	7 (2.4)	5 (1.7)	2 (0.7)

Figures courtesy of the Chrysler Historical Collection. Reprinted from *Mopar Performance News*.
**Figures not available.

WEDGE WARRIOR

Building Chrysler's Stump-pulling 440 Wedge

By Bruce Hampson

It's easy to overlook the 440 Chrysler motor. Buried deep in the bowels of motor homes, four-door land yachts, and police cruisers, the maligned Pentastar Wedge is neglected by many Mopar fanatics when contemplating performance in a serious vein. Despite the 440's inferiority complex, however, the Chrysler big-block can present the Mopar street runner with a rock-steady and powerful alternative to the traditional Ford and Chevrolet "glamour engines"—at substantially less expense.

It's been better than six years since the last factory-fresh 440 came down the pike; twice that since Chrysler offered the mighty Wedge in any semblance of true high-performance trim. Even so, nearly a million of the raised-deck "RB" motors

saw extensive street duty during its 13-year reign, and there remains an abundance of early Mopar muscle in every boneyard between Hollywood and Hoboken. There's never been any shortage of power-inducing componentry for the mill, either, compliments of Direct Connection (Chrysler's high-performance parts division). But all other qualifiers pale in comparison to the 440's ease of assembly. Relatively speaking, it's like bolting together a small-block Chevy—on a grander scale.

Since the 440 was available in a multitude of power ratings during its existence (all the way up to 390 stock ponies), we felt there wouldn't be much of a problem coaxing a few more from the bowels of the largest displacement engine in Chrysler history. Five hundred horsepower was

the target—but then again, so was the street. With 440 inches to work with, finding 500 hoof-stomping Clydesdales isn't hard; finding 500 that will be around next week is, especially when operating within the restrictive parameters imposed by today's impotent pump fuels. Chrysler employed 13½:1 compression in their killer street motors back in the Sixties; we were looking for even greater results while keeping a 9.5:1 ceiling on the squeeze.

Alchemist for this buildup was Bill Bagshaw, a name synonymous with Mopar performance throughout the early Seventies. The knowledge Bagshaw gained by throttling Pentastar big-blocks in the Pro Stock ranks has since been funnelled into Pro Parts, his Chrysler performance em-

The mill from the Imperial was obviously well cared for. While you never know what you're getting with a "rebuildable" non-running engine, ours showed extremely little wear. A check of the main bearing bore revealed surprisingly straight alignment, reinforced by the excellent condition of the forged crankshaft (cast cranks were not introduced in the 440 until 1974). The crank was magnafluxed and micro-polished.

Bagshaw's racing background is obvious in his fastidious approach to block preparation. The block and heads were subjected to an acid bath rather than a hot tank, then Bill scoured out every orifice with solvent and a wire brush. The motor was also pressure checked prior to the final rebuild.

porium. Bill's intent with the 440 was to construct a durable powerplant capable of resisting the stress of a street-and-strip motor. Reliability, economy, and power were Bill's main goals—in that order. As a result, there's nothing trick to this mill, no frills, and nothing not totally in keeping with his requirements for continual operation on the street (with frequent side trips into the nether regions of performance).

The engine itself was scavenged from the frame of a '68 Imperial. The preponderance of big-blocks in Chrysler's luxury fleet makes salvage operations quite easy, but you have to know what you're looking for. Any performance engine is only as strong as its weakest link, and all 1976 and newer Wedges are of the thin-wall variety unsuitable for anything greater than a .020-overbore.

Since our 440 was classified as a non-running "rebuildable" motor (it cost $100; functional used mills will set you back $250 and up), we had no idea what lay beneath the years of grease it wore like a shroud. After tearing the engine apart and subjecting the block and heads to an acid bath, however, we were quite surprised to discover minimal wear on all the major components. The block itself revealed cylinder bores needing just a .005-inch overbore to ensure proper piston travel. In keeping with the minimal oversize pistons available for the powerplant, though, the cylinders were bored an additional .030-inch. The block itself was decked slightly (.004-inch), and Bagshaw also checked for proper main bearing bore alignment.

The original crankshaft was in surprisingly good condition as well; after magnafluxing for stress cracks, the forged unit was polished prior to going back into the engine (cast cranks weren't introduced in the 440 until 1974—another reason for finding an early motor). The stock rods were also magnafluxed, then shot-peened and re-sized. Bill opted for high-strength Direct Connection ⅜-inch steel rod bolts and nuts (PN P4120068). At the other end of the original heavy-duty rods, Bagshaw decided upon a set of flat-top .030-over TRW forged pistons and rings.

After submitting no less than five different cylinder head designs to extensive flow bench testing, Bill settled on a modified version of Chrysler's standard head common to '68-70 big-blocks (casting no. 2843906). While the head features excellent flow characteristics, Bagshaw was able to improve upon the unit's performance tremendously by utilizing a DC porting template (PN P4120437) as a guide in opening up the bowl area between the valve seat and the radius on the port floor. Since oversize valves were also installed in the heads (2.18-inch intake, 1.81-inch exhaust), keep in mind that the template will have to be enlarged accord-

The stock connecting rod (right) is fine for most street/strip applications. The original rods were magnafluxed, polished, shot-peened, and re-sized. Since the connecting rod bolt is the weakest link in the bottom end, Bill opted for a set of high-strength DC ⅜-inch steel rod bolts and nuts (PN P4120068). The rod on left is a DC forged steel unit.

The biggest restriction to oil flow on the 440 is the stock ⅜-inch diameter pickup pipe. Chrysler recognized this deficiency by fitting the Hemi with a ½-inch tube; you can achieve the same results by drilling out the hole in the block where the pickup screws in, tapping it with ½-inch pipe threads, and employing a larger Milodon pickup (shown to right of stock unit).

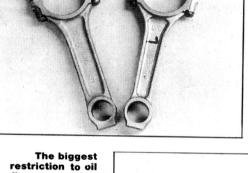

Nothing ever comes out perfect, so it's a good idea to mike all the pistons and over-bored cylinders, matching those closest in size. The pistons themselves, listed as replacements for '70-72 440 Six-Pack engines, create 9.6:1 squeeze when used with the 86cc head and .018-inch gaskets.

Direct Connection's Performance Book is an invaluable aid to anyone attempting a Mopar buildup. In fact, Bagshaw checked the DC recommendations as he assembled the 440 and discovered only one exception, concerning valve springs. The spring recommended (seen center: PN P3690933) required machining the heads; instead, Bill used a spring actually advocated for the smaller "A" blocks (right: PN P3412068), with no machine work needed. The stock spring is to the left.

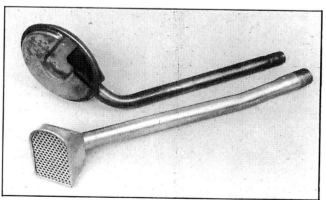

ingly. If you use a template, be sure to stop where indicated; removing excess metal from the heads will negate any improvements. The heads were also modified to provide a smooth transition between head and block, accomplished by fitting the head to the block, scribing a line around the combustion chamber from below, and blending the chamber into this line. Finally, the valve guides in the heads were bronze-walled and machined to ac-

cept DC seals, and the heads were milled to assure a fresh, flat gasket surface.

The valvetrain is a combination of Direct Connection, Crane, and TRW components. The cam itself is right from the DC catalog (PN P4120237). The biggest hydraulic grind DC offers, it spec's out at .509-inch lift, 292 degrees duration. However, since the flow bench illustrated optimum air flow for the 906 heads at between .400- to .420-inch lift, any of the four other hydraulic cams listed in the DC Performance Book can be substituted

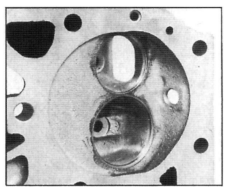

The 440 is blessed with free-breathing cylinder heads. Even so, a bit of porting can work wonders—and provide tremendous horsepower gains. Using a DC porting template (PN P4120437) as a guide, the bowl area between the valve seat and the radius of the port floor was opened up. Used in conjunction with oversize valves (2.14-inch intake; 1.81-inch exhaust), the modified 906 heads flowed as much air as either the Stage IV or V performance heads through .420-inch valve lift.

The 440 bolt-ons include a Holley Dominator intake manifold and 750 cfm double-pumper carb, DC mechanical tach drive ignition conversion kit, Pro Parts electric water pump drive kit, and new, blue anodized featherweight DC aluminum valve covers. The Holley intake/carb combination is the optimum induction system for this engine.

THE WINDS OF CHANGE

Air flow is critical to the performance of any naturally aspirated engine, and while the Chrysler 440 is blessed with free-breathing cylinder heads, they are not without normal production compromises inherent to all factory hardware. Much of the success of the 500 horsepower street Wedge is attributable to data acquired on the flow bench: Bagshaw subjected five different head castings (in a variety of alterations) to extensive flow testing before deciding upon a slightly modified version of the readily available (and inexpensive) 906 casting. We can't list the results of every head tested in every configuration, but some of the facts uncovered are enlightening, to say the least.

Employing oversize (2.14-inch intake; 1.81-inch exhaust) valves in moderately ported castings, for example, Bill found the 906 head to equal the flow ratings of either of Chrysler's premium Stage IV or V parts up through .420 valve lift—excellent range for a street killer. Intake flow in the 906 began leveling off at that point, though exhaust flow continued through .500-inch lift. Speaking of exhaust flow, one of Bagshaw's surprising highlights was tracing the flow of the 906—versus the vaunted Stage V—during air expulsion; in stock form, the 906 actually flowed greater than the Stage V! Bill ran 15 different tests, all told, from stock to radically ported units. The flow information (which can be acquired from Pro Parts for the miserly sum of five bucks) is an invaluable tool for anyone contemplating a 440 buildup.

without serious loss of power. The cam was installed on a 108-degree intake centerline and required the use of a five-degree advance offset button.

A set of DC hydraulic rocker arms on TRW rocker arm shafts (using Crane steel retainers and locks) actuate the oversize valves. Bagshaw found that the best valve spring for the engine was in fact recommended as a replacement part in smaller "A" blocks (PN 3412068). The spring originally called for required the machining of the inner spring set and shortening the dampener between the inner and outer coil spring so that it wouldn't coil bind! The installed height of the substituted unit is 1.725 inches, yielding 360 pounds open pressure at .500-inch-lift. The real benefit, was in skipping the machine work.

Clean, easy, and effective. The perfect combination.

If anything, Bagshaw used the 440 as a testbed for verifying much of the information found in the Direct Connection Performance Book—and discovered, with the lone exception of the valve springs, that it was right on the money. Bill completed the powerplant with the addition of a DC mechanical tach drive ignition conversion kit (PN P4286512), DC 8mm plug wires, a Pro Parts electric water pump drive kit, and Milodon oil system. A box-stock Holley Street Dominator intake manifold and Holley 750 cfm double-pumper capped it all off.

The only question was, would it run . . . and how hard? Would you believe

511 horsepower at 5500 rpm?! The only changes made to the engine once hooked to the dyno was to block off the heat riser in the intake manifold; that alone was worth 26 horsepower. Even more important from the street angle, however, are the torque readings. Optimum torque (and with it, maximum fuel efficiency) came in at just a thousand rpm lower; 522 lbs./ft. at 4600 rpm. There's enough stump-pulling power in this mill to fell a redwood!

What's more, it's relatively docile. And, it can be repeated *ad infinitum.* The parts are all either stock (and on a 440, that translates into heavy duty), Direct Connection, Crane, or TRW; all proven performers in one cohesive package. And that's the key. The entire package, by the way, is available directly from Bagshaw's Pro Parts outlet. Remember Bagshaw's goals? You should; he met or exceeded every one of them. Reliability. Economy. And *Power.* 🅖

Bagshaw's homework in crafting the optimum street Wedge has been assembled in one package, offered by Pro Parts. For $1195, the Pro Parts "Super Street" kit includes everything necessary to build a 500-horse street 440 (you provide the block, heads, crank, rods, and induction and ignition systems).

During the dyno run, Bagshaw blocked off the heat riser in the intake manifold with this machined aluminum plate; DC offers a package (PN P4120476) that accomplishes the same thing by incorporating a stainless steel manifold block-off at the gasket. The dyno revealed this one change to be 26 horsepower!

Pro Parts' Kim Welch and Bill Bagshaw ready the Wedge warrior for the moment of truth. Using a box-stock carburetor, the 440 produced 511 horsepower at 5500 rpm, and torque readings of 522 lbs./ft. at 4600 revolutions per minute!

SOURCES

Crane Cams
100 N.W. 9th Terrace
P.O. Box 160
Hallandale, FL 33009
305/457-8888

Direct Connection
P.O. Box 1718
CIMS 423-13-06
Detroit, MI 48288
313/497-1220

Holley Replacement Division
11955 E. Nine Mile Road
Warren, MI 48090
313/497-4000

Pro Parts
4113 Redwood
Los Angeles, CA 90066
213/821-2911

TRW Replacement Parts Division
8001 E. Pleasant Valley Road
Cleveland, OH 44131-5582
216/447-8164

DYNO RESULTS

ENGINE RPM	HORSEPOWER	TORQUE
3000	255	448
3500	323	483
4000	391	509
4500	453	522
5000	487	514
5500	511	481
6000	496	434

CYLINDER HEAD SWAP SHOP

While there are some major variations between Mopar low-block (361-383-400) and high-block (413-426-440) big-inch Wedge motors, none of the differences concern cylinder head interchangeability—and Chrysler produced well over a dozen separate head designs during the combined 20-year reign of the engine series. We've assembled a listing of the more plentiful castings, some of which are still available from Direct Connection (as are the Stage IV and V heads). With some four million "B" and "RB" engines manufactured, however, a few salvage operations at the local boneyard should be adequate for finding the right heads for your particular Mopar.

Casting Number	Application	Features
2406516	'64-67, 361-383	1.60-inch exh. valves
2780915	'67 440, hi-perf.	1.74-inch exh. valves
2843906	'68-70, 383-400	'67 head w/lrg. comb. cham.
3462345	'71-72, 383-400-440	Emission head
3462345	'73, 400-440	
3751213	'73, 400-440	Motor home head
3769902	'74, 400-440	
3769975	'75, 400-440	
4006452	'76-78, 400-440	

MOPAR SMALL-BLOCK I.D. GUIDE

BY DEV ANAND

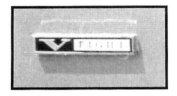

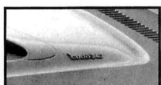

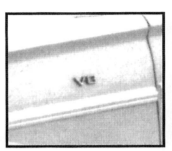

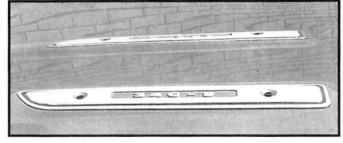

A Detailed Look At Chrysler's Little Engine That Could

The Chrysler small-block, known as the "LA" engine, has been offered in displacements of 273, 318, 340, and 360 cubic inches. Dating back to 1964, it is simply one of the best V-8s ever designed. A product of lightweight casting techniques, the "LA" small-block offers legendary durability, performance, and reliabilty.

The "LA" engine family reached the peak of development in 1970 with the introduction of the 340 Six Pack. With three Holley deuces on an Edelbrock high-rise manifold, it was just what those free-breathing 340 cylinder heads needed. It put the slightly overweight "E" body 'Cudas and Challengers into the 14's, right off the showroom floor!

In 1974, most of the high-performance 340 four-barrel pieces were bolted on the longer-stroke 360, again allowing the small-block to shine. In 1978, the 360 four-barrel hi-po pieces were once again installed, this time on a D150 truck. It was one of the quickest-accelerating domestic vehicles that year!

One of the advantages of the "LA" small-block is their parts interchangeability. In most cases, intake manifolds, cylinder heads, connecting rods, and crankshafts can all be interchanged.

In the following pages, join us in taking a look at what makes the "LA" small-block such a street stormer. 🄶

"A" engine block is a lightweight (160 pounds with main caps) thinwall casting with 4.46-inch cylinder bore center spacing.

High-performance versions of the 273 appeared in 1966 models. The engine featured a Carter AFB carb, longer-duration camshaft, domed 10.5:1 pistons, and factory power ratings of 235 hp at 5200, and 280 lbs-ft of torque at 4000. Note the 90-degree oil filter mounting used on pre-1972 small-blocks.

The 1971 and later "LA" small-blocks used spread-bore Thermo-Quad carburetors. The cast iron dual-plane high-rise intake manifolds used with these carbs are hard to improve on.

This early 273 block shows the two-bolt main caps that were on all production "LA" engines, along with oil pump mounting pad machined into the rear main cap.

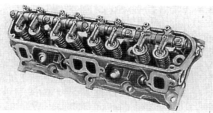

This early 273 cylinder head (casting number 2465315) featured 1.78-inch intake and 1.50-inch exhaust valves. Note that the mechanical valve gear used on all 273s from 1964-67 "LA" cylinder heads weigh about 41 pounds each.

The "LA" oil pump (left) bolts to the rear main cap (center). Unlike Chevrolets, Mopar oil pick-up threads into the pump, eliminating the worry of a pressed-in pick-up vibrating out.

All production "LA" pistons are of cast construction. The 273, pre-'73 318, and 340 pistons are all full-floating design. The 360 piston and pin (shown) was the first to use a pressed design.

340-V8

BORE – 4.04
STROKE – 3.31

- TWO LEVEL INTAKE MANIFOLD 2.2 SQ. IN. BRANCH
- NEW CYLINDER HEAD 2.2 SQ. IN. INTAKE PORTS
- HIGH LOAD VALVE SPRINGS & DAMPERS
- N9Y SPARK PLUGS
- 2.02 DIA. INTAKE VALVES
- LOW RESTRICTION EXHAUST MANIFOLDS
- 2.25 DIA. EXHAUST PIPES
- 10.5:1 C.R. PISTONS FLOATING PISTON PINS
- FORGED STEEL, SHOT PEENED CRANKSHAFT
- MAIN BEARINGS ALUM #1-2-4 BABBITT #3-5

- SINGLE 4BBL CARB WITH AUTO CHOKE
- LOW RESTRICTION NON-SILENCED AIR CLEANER
- TORQUE FAN DRIVE
- DUAL BREAKER DISTR.
- 1.70 SQ. IN. EXH. PORTS
- 1.60 DIA. EXHAUST VALVES
- HYDRAULIC TAPPETS
- MANUAL TRANS. CAMSHAFT: 276-284-52
- AUTO. TRANS. CAMSHAFT: 268-276-44
- .5 SQ. IN. SHANK CONN. ROD
- TRI-METAL CONN. ROD BEARINGS
- HIGHER MAIN BEARING CAPS THICKER CYLINDER BLOCK BULKHEADS
- OIL PAN WINDAGE TRAY
- DOUBLE ROW ROLLER TIMING CHAIN

275 BHP @ 5000 RPM
340 LB FT @ 3200 RPM

The introduction of the 340 in 1968 set the performance world on its ear. Installed in Darts and Barracudas, this conservatively rated 275-horse small-block gobbled up big-blocks of all makes.

For More Info

Any street machiner contemplating the slightest hop-up of a Chrysler should have a copy of Mopar Performance's book *Engine Speed Secrets* (PN P4349340). This is the bible of Mopar street machiners and contains much of the data presented here. You can order it from any Chrysler dealer or Mopar Performance outlet.

340 Six Pack Major Parts

Part numbers are for reference only. They are no longer available new.

Trans-Am block	3577228
Cylinder head	3577037
Right intake rocker arm	3577076
Left intake rocker arm	3577077
Exhaust rocker arm	3577078
Rocker shaft	3577086
Pushrod	3577121
Intake manifold	3418681
Front carb	3577184
Center carb (auto trans.)	3577183
Rear carb	3577185

Mopar crankshafts are legendary in their beefiness, and the "LA" engine is no exception. Chrysler first used cast cranks in 1971 with the 360. The 360 crank (shown) also has larger main bearing diameters (2.81 inches). The 273, 318, and 340 engines all used 2.50-inch main journals. Thrust is taken by #3 main journal.

Years Of Production In Passenger Cars By Cubic Inches

273cid: 1964-1969
318cid: 1967-present
340cid: 1968-1973
360cid: 1971-1980
(Note: The 360 engine is still in production in trucks.)

Those looking for small-block cylinder heads should search for casting number 3418915. This casting is found on the dirt-cheap 1971-72 360 two-barrels and is identical to the high-flow 340 heads except for intake valve size. The same casting was used on the 1972 340 and (with revised pushrod holes) 1970 340 six-barrel.

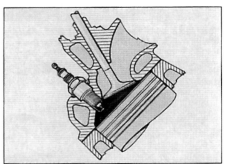

All "LA" cylinder heads have two features that aid their high efficiency. The spark plug location is at the very top of the chamber, just like the Hemi. And to allow bigger valves for a given bore size, the valves open on the centerline of the cylinder in side view.

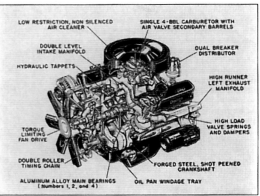

This cutaway shows the structure of the 340. Forged crank, dual-point distributor, and high-rise intake manifold all contributed to making this engine a street and strip legend.

Production Bore and Stroke

3.63 × 3.31 = 273 cid
3.91 × 3.31 = 318 cid
4.04 × 3.31 = 340 cid
4.00 × 3.58 = 360 cid

Connecting Rod Bearing Journal Sizes, Lengths, And Bolt Diameters

All LA engine rods center-to-center length: 6.123 inches
All LA engine rod bolt diameter: ⅜ inch
All LA engine rod journal size: 2.125 inches

The 1966 275-horse 273 "D-Dart" would turn 14-second quarter miles. Note the Holley 4160, List 3778 carb, complete with manual choke and carb spacer! Headers are by Doug. D-Dart was not the most tractable thing in traffic.

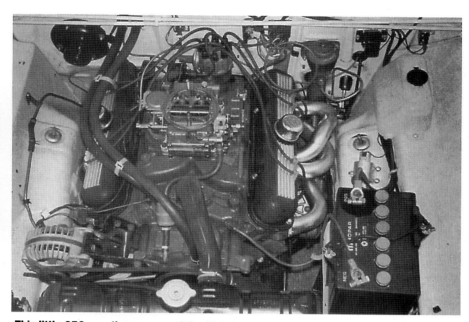

This little 273 was the scourge of the highways in 1966. Known as the "D-Dart," (for the D/Stock racing class) this special 273 four-barrel was rated at 275 horsepower. Doug's headers, big Holley four-barrel, a solid-lifter 284 duration ,510-inch-lift cam-shaft, and heavy-duty valvesprings were also part of the package.

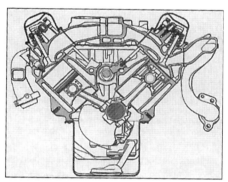

All small-block "A"-body exhaust mani-folds have very different designs, passen-ger side versus driver side, due to the steering box location. This 273 uses a three-branch, low-runner design with a center outlet on the passenger side. The driver's side has a high runner that ex-tends to the rear of the engine before de-scending into an S-shape to clear the steering column.

The 340 and 360 engines were among the first to receive Chrysler's new Elec-tronic Ignition System in 1972. While many Chrysler engines could no longer take advantage of its smooth, reliable per-formance to 6000 rpm, street machiners (and Mopar Performance) soon learned to adapt it to the older, higher-revving en-gines.

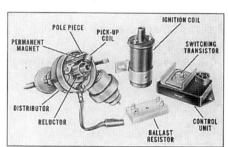

The slotted distributor drive gear slot points towards the front left intake mani-fold hold-down bolt at TDC. Note the square-flange 340 manifold; this identifies a 1968-70 intake designed for a Carter AVS.

A workhorse engine, the early 360 two-barrel is good junkyard pickings for low-buck performance. The cylinder heads are the same as the early 340s, except for in-take valve size. This '73 example has air injection, however, which used different (read more restrictive) castings.

The main difference between the 1972 and 1973 340 engines was the crank. Up until 1972, the 340 had a forged crank, and used the balancer on the right. The 1973 cast crank engines were externally balanced, requiring specially weighted ba-lancers (left).

Engine Swapping

All 273, 318, 340, and 360 engines use the same bellhousing bolt pattern and driver's side mounting lugs. The 340/360 passenger side mounting lugs are configured differently from the 273/318, thus requiring the appropriate motor mount bracket for engine swapping.

Factory Engine Colors

Data is according to best information, some exceptions exist.
273cid—1964-69: Red
318cid—1967-72: Red or Blue, 1972-83: Blue, 1983-present: Black
340cid—1968-69: Red, 1970-71: Street Hemi Orange, 1972-73: Blue
360cid 1971-83: Blue, 1983-present: Black

"LA" Dimension Chart*

* All measurements in inches

Width	
Valve cover to valve cover	20.5
Oil pan	8.25
Exhaust Manifolds (340)	27.1
Length	
Block	22
Fan belt to rear of block	27
Fan to rear of block	31.8
Height	
Pan bottom to air cleaner top (340)	29.0
Pan bottom to carb top	26.5
Pan bottom to valve cover top	22.0
Pan rail to valve cover top	14.5

MOPAR SMALL-BLOCK I.D. GUIDE

Underneath all those hoses is the power-plant for one of the quickest-accelerating domestic vehicles in 1978. It is a special high-performance 360 four-barrel, dropped into a Dodge D150 half-ton truck! The fire engine red "Li'l Red Express Truck" would run e.t.s in the 15's.

"LA" Cylinder Head Casting Numbers

Year	Engine	Casting #
1964-65	273	2465315
1966	273	2536178
1967	273/318	2658920
1968-71	318	2843675
1968-71	340	2531894
1970	340 6-bbl	3418915
1971	360	3418915
1972	318	2843675
1972	340/360	3418915
1973-74	318	2843675
1973-74	340/360	3671587
1973	360 w/air pump	3671587
1975	318	3769973
1975	360	3769974
1976	318	3769973
1976	360	3671587, 3769972
1977-83	318 std.	4027163, 4027593
1977-80	360	4027596, 4071051
1981-83	318 HP	4027596, 4071051

1978 360cid Li'l Red Truck

In the late '70s, Chrysler engineers took a D-150 pickup and fitted it with many special parts from the old 340. Painted fire engine red and dubbed the Li'l Red Truck, this truck vied for the honors as the fastest-accelerating American vehicle in the 1978 model year! The following are some of the pieces added to the standard 360 four-barrel to make the 360 Express engine:

•Camshaft from the 1968 340 4-bbl.
•Red stripe valvesprings with damper from 1968 340 4-bbl.
•Standard valve retainers (replacing rotators)
•Large Thermo-Quad from Police 360
•Intake manifold from 1978 Police 360
•Windage tray from Police 360
•Roller timing chain and sprockets from Police 360
•Dual-snorkel air cleaner with fresh air ducts to radiator yoke
•Chrome valve covers and air cleaner housing lid
•Street-Hemi style mufflers

Crankshaft Material And Weight

All 273cid: Forged steel
1967-1972 318cid: Forged steel
1968-1972 340cid: Forged steel, shot-peened
1973-present 318cid: Nodular iron
1973 340cid: Nodular iron
All 360cid: Nodular iron, 360 hp are also shot-peened
Production LA cranks weigh approximately 55 pounds

The ultimate small-block Mopar was the 340 Six Pack that was standard on the 1970 AAR 'Cuda and Challenger T/A. Main features were a special block that would accommodate four-bolt main bearings, intake pushrod holes relocated to allow radical porting, special high-strength valve gear, and famous Six-Barrel induction system.

Crankshaft Main Bearing Sizes

All 273, 318, 340cid: 2.50 inches
All 360cid: 2.81 inches

"LA" Engine Trivia Quiz

Q: Which production "LA" engine intake manifold is not physically interchangeable with the rest, and what year or years was it offered? **A:** *The 1964 and 1965 273 cylinder head had the intake bolt holes drilled at a unique angle. These heads and intakes must be used together.*

Q: According to the factory, what is the maximum allowable overbore of an "LA" engine? **A:** *0.040-inch*

Q: In 1972, all "LA" engine vibration dampers were revised. What was this change? **A:** *1972 and later vibration dampers had their pulley bolt pattern changed from an offset to a symmetrical layout, requiring the older pulleys to be used with the older dampers.*

Q: If the distributor is removed out of an otherwise correctly running "LA" engine, how many different ways could it be reinstalled? **A:** *Two. "LA" engines have a slotted intermediate shaft that meshes with the cam gear. The distributor can only be installed correctly, or 180-degrees out.*

Q: What is different about the oil filter in most 1972 and later "LA" engines? **A:** *The position of the oil filter was changed from angling back to the rear of the car to sticking straight out from the side of the block.*

Q: Can all "LA" engines use the same oil pan? **A:** *No. The 360 uses a special pan because of its larger main bearing diameters.*

Q: What was the first "LA" engine to be fitted with a windage tray, and what model year was it? **A:** *The 340cid, in 1968.*

Q: What change was made to the water pump in 1969-and-later passenger car "LA" engines? **A:** *The water pump inlet (lower hose) was changed from the passenger's side to the driver's side.*

Q: Did any "LA" engines come with an aluminum intake manifold as standard, and if so, what engine size(s) and year(s)? **A:** *Yes. The 1970 340 Six Pack used an aluminum Edelbrock intake manifold.*

Q: What is the primary difference between the early 340 cylinder head and the early 360 head? **A:** *The 340 has a larger intake valve size.*

Q: What was the factory horsepower rating of the 1968-1971 340 four-barrel? **A:** *A very conservative 275.*

MOPAR BIG-BLOCK
I.D. GUIDE

BY DEV ANAND

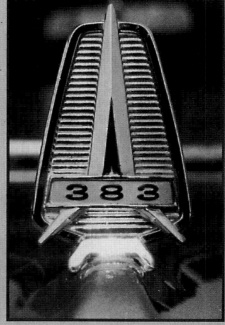

Although it has been out of production for over 10 years, the big-block Mopar is still one of the most feared and respected engines ever to rumble down a boulevard. In late-night bench racing sessions, legendary displacements are heard again and again: 383, 413, 426 Max Wedge, and 440 Six Pack.

The big-block Chrysler comes in two major flavors: the low-block "B," and the taller "RB" (Raised "B") engine. All "B" engines share a common

stroke and deck height, as do the "RB." They all share the same cylinder heads and engine front end parts. According to Chrysler, over three million 383s were built along with more than 750,000 440s.

Their legendary durability comes from typical Chrysler over-engineering. Anybody who has looked at Chrysler rods and cranks quickly realizes how bulletproof these engines are. With the exception of the Max Wedges, these are low-rpm (up to 6000 rpm) engines.

Face-distorting torque and monstrous mid-range are what make these engines so unbeatable on the street.

A staggering array of induction systems has graced the big-wedge head, some bordering on the bizarre. The early long-ram manifolds, the inline dual quad, the cross-ram Max Wedge, and the legendary Six Pack were all results of talented engineering.

Let's take an in-depth look at those fearsome Chryslers with the distributor at the front—where it belongs!

MOPAR BIG-BLOCK
I.D. GUIDE

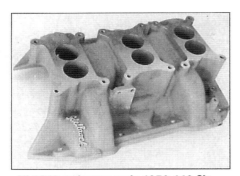

All 1969 and some early 1970 440 Six Pack manifolds were aluminum and made by Edelbrock. These original manifolds are ID'd by the "CHRY 3412048" between the front and middle carb flanges. All subsequent Six Pack Chrysler manifolds were cast iron.

Arrows show the enlarged primary throttle bores of the Dodge Ramcharger (Max Wedge) 426-A. The primary bores have been widened ¼ inch, matching the 1¹¹/₁₆-inch secondary bores and larger primary throttle blades. The aluminum cross ram uses threaded plugs for access to the intake manifold bolts.

This 1958 Coronet Custom Cruiser CHP patrol car turned 17.3 at 84 mph powered by a special-order 320-horse dual-quad 361cid "B" engine. The base single four-barrel engine had a rating of 305 horse-power.

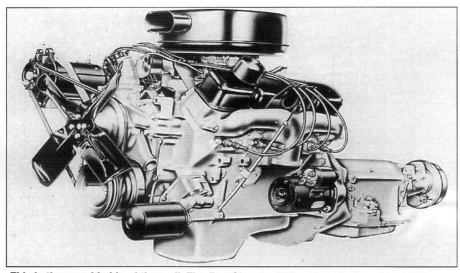

This is the granddaddy of them all. The first Chrysler "B" engine displaced just 350 cubic inches in 1958. Bore and stroke was 4.06 x 3.38. The high-performance D-500 model was a 361. Note the solenoid starter, center outlet exhaust manifolds, and oil bath air cleaner on this 350.

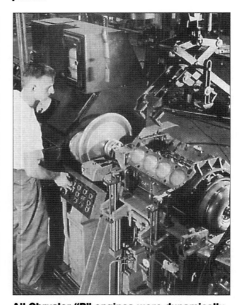

All Chrysler "B" engines were dynamically balanced at the factory. This is one of the first engines down the assembly line in mid-1957. Note the torque converter attached to the back.

Cylinder Bore Spacing And Weight
All "B" and "RB" engines have a cylinder-to-cylinder bore spacing of 4.80-inch. A "B" engine block weighs 225 pounds with main caps, while an "RB" is about 25 pounds heavier.

Years Of Production In Passenger Cars By Cubic Inches
350cid: 1958
361cid: 1958-1966
383cid RB: 1959-1961
383cid: 1959-1971
400cid: 1972-1978
413cid: 1959-1965
426cid: 1963-1965
440cid: 1966-1978

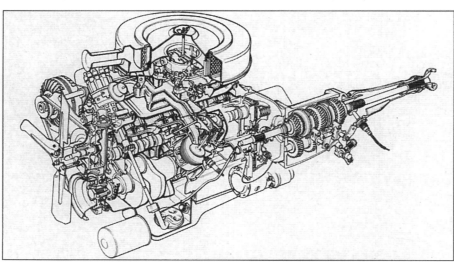

The 383 was the workhorse engine in the line-up for years. Note the simple and rugged construction, such as the shaft-mounted rocker arms, waterless intake manifold, integral coolant bypass, huge bearings, and deep skirt block.

This rare photo is a 1972 440 Six Pack configuration. Chrysler made few (if any) 1972 versions, a casualty of emissions and insurance. Note the absence of the hot air stove on the driver's side exhaust manifold.

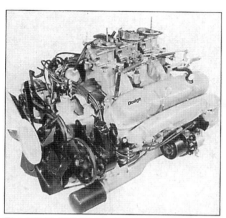

The 1970 440 Six Pack was simply the pinnacle of street development of the big-block. Smooth and docile in traffic, at wide open throttle 1355cfm of six-barrel induction would propel 4000-pound Dodges into the low 14's. And best of all, it had hydraulic lifters!

Paul Ackerman was Director of Engineering at Chrysler during the development of the new "B" engine. The deep skirt block, short stroke design, stamped rocker arms, and integral valve guides were all the latest rage and were incorporated into the design. The engine would thrive for 20 years.

The 361 Long Ram D-500 engines were the talk of Detroit in 1960. Its wild looks fit perfectly in that cold war era of Sputnik, fins, and chrome. The system was designed for maximum torque in low- to mid-rpm ranges.

The 400 four-barrel was introduced on the 1972 Plymouth Road Runner, featuring a Carter ThermoQuad and dual exhaust. The two-barrel versions used a Holley carb and single exhaust. The 400 replaced the 383 as the workhorse engine of the line-up.

This 1960 Chrysler 300-F sports a ram-tuned 375-horse 413. This 4700-pound boat streaked to a 16.0 at 85 mph quarter mile!

Boring
Blocks without coreshift cast before 1974 should be able to be bored 0.060-inch. Newer blocks are good for 0.030-inch.

Crankshaft Main Bearing Sizes
All "B" Engines: 2.625-inch
All "RB" Engines: 2.750-inch

Production Bore And Stroke
"B" Engines
$4.0625 \times 3.375 = 350$cid
$4.125 \times 3.375 = 361$cid
$4.250 \times 3.375 = 383$cid
$4.342 \times 3.375 = 400$cid

"RB" Engines
$4.0312 \times 3.750 = 383$cid
$4.1875 \times 3.750 = 413$cid
$4.2500 \times 3.750 = 426$cid
$4.3200 \times 3.750 = 440$cid

The mid-'69 introduction of the 440 Six Pack was on Dodge Coronet Super Bee. The engine was rated at 390 horses, with 490 lbs-ft of torque. The hood was a pinned, lift-off type!

MOPAR BIG-BLOCK I.D. GUIDE

The modern "RB" engine has a rugged but simple design. Its 250-pound cylinder blocks are all of a two-bolt design. The dual-plane cast iron manifold is made for low- and mid-range torque.

The 440 Magnum engine was introduced in 1967. Its 2.08-inch intake valves and 1.74-inch exhaust valves, free-breathing heads and manifolds, windage tray, and twin-snorkel air cleaner were major highlights of the 375-horsepower engine.

Crankshaft Material And Weight
All 350, 361, 413, and 426: Forged steel
All 383 up to 1971: Forged steel
All 440 up to 1974: Forged steel
Some 1972-74 400 four-barrels:
 Forged steel
Late 1971 383 two-barrels: Cast iron
Most 1972-78 400: Cast Iron
1974-1978 440: Cast Iron

"B" Engine Cylinder Heads

Year	Engine	Casting #
'60-62	361-383-413	2206324
'62	413 Max Wedge	2402286
'64-67	361-383	2406516
'64	426 Max Wedge	2406518
'63	361-383-413	2463200
'63	426 Max Wedge	2463209
'67	440 HP	2780915
'68-70	383-440	2843906
'71-72	383-400-440	3462346
'73	400-440	3462346
'73	400-440 M.H.	3751213
'74	400-440	3769902
'75	400-440	3769975
'76-78	400-440	4006452

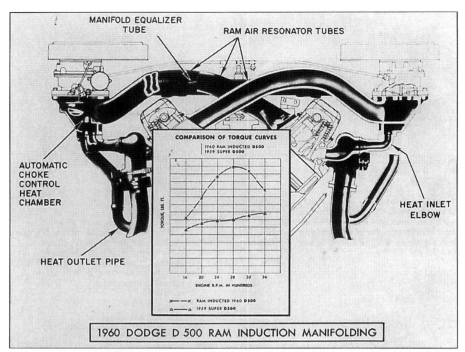

1960 DODGE D 500 RAM INDUCTION MANIFOLDING

The Ram induction manifolding was available from 1960 to 1962 on certain 361, 383, and 413 engines. Each Carter AFB fed the cylinders on the opposite bank, 30 inches away. The length tuned the torque peak at 2800 rpm. Their output dropped off quickly above 4800 rpm, however.

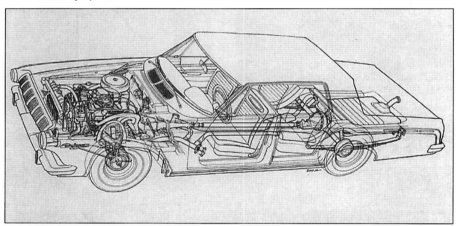

This cutaway shows the unique exhuast manifolds that were an integral part of the '63 426 Max Wedge (called Ramcharger by Dodge). Large, 3-inch collectors flared out to a bolt on "lakes" pipe, for quick uncapping at the strip! These engines and vehicles were sold "as is" and did not come with a warranty.

The year 1964 was final hurrah for the Max Wedge. It got new Tri-Y cast headers (they were heavy!), seven-blade fan with fluid drive, modified combustion chambers, bigger Carter AFBs, new high-lift camshaft, connecting rods, and crankshaft. Two compression ratios were available, 11:1 and 12.5:1.

MOPAR BIG-BLOCK I.D. GUIDE

Frank Walter, Plymouth's chief engineer displays the new 413 Max Wedge engine. Designed to compete in Super Stock, it would quickly become the powerplant to beat.

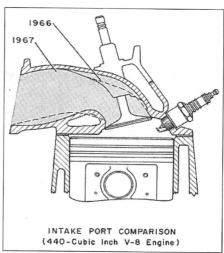

INTAKE PORT COMPARISON
(440-Cubic Inch V-8 Engine)

The major improvement in big-block street heads occured in 1967. A redesigned port, along with big valves, made these heads the ones to look for. The 1968-70 heads were similar, but had a much larger combustion chamber for use with increased deck height.

The Carter Thermo-Quad was made standard on 440 engines in 1973. The resin bowl is designed to insulate the fuel from heat, keeping it more than 20 degrees cooler. Properly tuned, these were excellent-responding carburetors, with lots of breathing capacity.

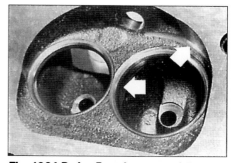

The 1964 Dodge Ramcharger (and Plymouth Super Stock) engine had their combustion chambers modified to improve air flow. Machined reliefs (arrows) in the sides of the chambers adjacent to the intake valve and the entrance to the intake valve seat were both done. The '63 head had no machining in these areas.

Big-Block Trivia Quiz

Q: The "B" engine is known for its ruggedness. How many head bolts does it use per side? **A:** *17*

Q: What was special about the valve gear on pre-1964 engines? **A:** *These early engines did not have a cast-in rocker arm support or stand. The support and rocker shaft were attached by long bolts.*

Q: There is a way to tell at a glance if a "B" or "RB" engine is newer than 1964. What is it? **A:** *The valve cover bolt pattern. In 1964, Chrysler added two additional hold-down screws at both lower outboard edge of the covers. Thus, late-model covers have six bolts each.*

Q: Did any "B" engines come with an aluminum intake manifold, and if so, what were they? **A:** *Many did. The 1960-62 "Long Ram" engines, the '62-64 Max Wedge engines, and the 1969 440 Six Pack all came with aluminum manifolds.*

Q: "RB" and Hemi cranks differ externally in just one major way. What is it? **A:** *"B" and "RB" Wedge engine cranks have a six-bolt flange. Hemi cranks are eight-bolt.*

Q: Why is the 440 Six Pack externally balanced? **A:** *Because of the heavy rods and pistons used in this engine.*

Q: What was unique about the blocks used in the 383 A-Body? **A:** *Most of the driver's side engine-mounting lug on most blocks was milled off. Two special bosses were used instead to mount the engine on that side.*

Q: What was special about the cam in the 440 Six Pack? **A:** *The lobe face is ground with a much larger radius (low taper) than the standard camshaft. This requires that special lifters be used with these cams. The actual lift and duration were the same as the camshaft in the 440 Magnum.*

Q: If the distributor is removed out of a correctly running "B" engine, how many ways can it be reinstalled? **A:** *Two. "B" engines have a slotted intermediate shaft that meshes with the cam gear. The distributor shaft can only be installed correctly, or 180 degrees out.*

Q: Can all "B" engines use the same oil pan? **A:** *Yes, with the appropriate pick-up tube.*

Q: The oil pump in your "B" engine just picked up some dirt and broke a gear. Now you have to pull the oil pan. True or False? **A:** *False. The oil pump is mounted externally at the left front of the engine.*

Q: Why is the 1967 440-hp cylinder head so sought after? **A:** *It had all the good porting of the "906" castings, combined with a small combustion chamber volume.*

Q: Can you use a 383 distributor in a 440? **A:** *Although they look the same, the 383 distributor shaft is slightly shorter and will not fit. The 440 distributor will work in a 383, if you have a small adapter spacer available from Mopar Performance.*

For More Info

Any street machiner contemplating the slightest hop-up of a Chrysler should have a copy of Mopar Performance's *Engine Speed Secrets* book (PN P4349340). Written by staff engineer Larry Shepard, this is the bible of Mopar street machiners, and contains much of the data presented here. You can order it from any Chrysler dealer or Mopar Performance outlet.

THE W2 CONN

A Mopar motor that runs well into the nines, with a single 4-barrel: it has to be a Hemi, or at least a 440, right? In years past, whenever long-time Chrysler racer Paul Rossi wanted to go fast, that's what he built. The 440's, especially, were dependable and affordable Super Gas motors ("Magnum Mirada, Part 2," HRM May '81; "Part 3," HRM June '81). But times change. Mopar no longer makes 440 motors, or the chassis to receive them—but they still offer inexpensive small-block (A-motor) components. The W2 race heads for the Chrysler small-block are the equal of the best small-block Chevy castings, yet can be had for only $250 a pair. Best of all, they'll bolt on to the millions of blocks littering the junkyards. So, with big-block pieces starting to dry up, when it came time to build a new Super Gas car for 1984, Paul decided to "get with the 1980s" and go with the smaller motor. Initially he laid grandiose plans for trick, tunnel-

ram-equipped engines with max-ported heads, billet cranks, and special race blocks. After all, reliable big-blocks ruled the faster bracket classes, and the little A-motor would be giving away lots of cubes. How could it ever hope to compete against the prickly 500-inch porcupines run by the competition?

Obviously, a full-scale development program was warranted. Initially, Paul put together a low-buck, single-carb, cast-crank "mule motor" as a learning experience, intending for it to merely run 10.90's and garner data for the more sophisticated engines to come. The results astonished the long-time B-motor proponent, with the junkyard motor running consistently in the nines! Now, for the time being, there is no need for a trick motor—the junkyard special gets the job done just fine. Sound incredible? When you see the "econo" nature of Paul's latest go-fast "recipe motor," you'll be even more impressed.

nals will be turned down anyway to the 2.0-inch size of an early small-block Chevy. In the process, the journals are offset ground to increase the stroke by .104-inch. With the bore now 4.050 and the stroke up to 3.685 inches, a total of 382 cubic inches results.

Normally, stroker motors require custom pistons and rods, but Rossi has come up with a relatively inexpensive (by racing standards) solution: Chrysler Direct Connection (DC) once carried a bunch of 355-inch NASCAR pistons made by Diamond Racing. Since the 355 Mopar isn't popular in NASCAR, Chrysler racers like Herb McCandless have ample supplies lying around gathering dust.

For this motor you'll need the .010-over versions, as shown in the Parts List. Because the 355 was a special short-stroke race engine, to work with the stroked 360 crank and achieve the desired 12.6:1 compression ratio (CR), the pistons must be machined so their domes and decks protrude only .150 and .045-inch, respectively, above the block's deck. To achieve this on a stock junkyard block will typically require a .250-inch dome whack, while .020-inch must be taken off the flat. Machining the pistons in this manner will save money, since the heads and block

Bottom-End

The plan is to start with a well-used (and hence, naturally stress-relieved) 360 block. Nobody wants them because they're "only" smog motors, with

low compression and cast cranks. Use a pre-'76 engine, since they have walls thick enough to accept a .050-inch overbore. If the crank is lunched, so much the better, because the rod jour-

BUILD A CHRYSLER SMALL-BLOCK THAT RUNS CONSISTENT NINES WITH A SINGLE 4-BARREL

By Marlan Davis

won't have to be milled. The piston cut has the added virtue of shaving additional weight off the already lightweight slug, while helping to improve flame propagation.

Stock 6.123-inch center-to-center length rods can be used with the 355 pistons. Rossi prefers Bill Miller or Venolia aluminum rods, since they both cut reciprocating weight even further, and are cheaper than fully-prepped steel rods. The big ends are machined for the Chevy 2.0-inch rod journals, while the small ends accept readily available, lightweight, .090-inch wall thickness, .990-inch o.d., big-block Chevy pins. It's a simple operation to ream out the .984-inch i.d. 355 Chrysler piston pinholes to accept the larger Chevy pins.

The turned-down crank, combined with the lightweight rods and pistons, allows the engine to rev quicker, offers less frictional horsepower losses, and permits the 360-based engine to be internally balanced. Internally balancing the motor in turn permits the use of

Old Chrysler 355 NASCAR pistons are an economical yet lightweight and high-quality alternative to costly custom-made setups. For use in Rossi's stroker motor, most of the dome must be machined off (right).

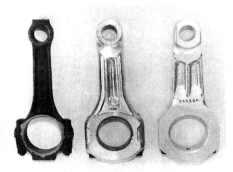

Shed reciprocating weight, gain horsepower. Besides, quality aluminum rods like these made by Bill Miller and Venolia are lots cheaper than fully prepped steel rods.

standard (and lighter) 318 or 340 harmonic balancers, flexplates, and torque converters.

When it comes to bearings, on the mains Rossi uses the new DC Clevite 77 360 main bearing set (see Parts List), which comes fully grooved for improved oiling. Early 283 Chevy rod bearings are required with the reground crank and aluminum rods. To fit the bill here Rossi selected a Sealed Power babbit-type "micro" set, as he feels such "softer" bearings are more forgiving in a drag engine than the Clevites, especially with a cast crank.

Rossi chose Speed-Pro for the rings. The "gap-to-fit" plasma-moly set contains conventional-design 1/16-inch moly top, 1/16-inch ductile iron second, and 3/16-inch low tension oil rings. While normal gaps are run on the first ring, the second is run extremely tight to achieve optimum sealing, as it runs cooler and doesn't expand as much.

Oiling

Since this engine uses a roller cam, the right and left-hand oil galleys must be blocked off to prevent lifter ejection-induced oil pressure loss. (Also, some roller lifter designs cause oil pressure loss by uncovering the oil galley.) Since the right oil galley also feeds the main, rod, and cam journals, typical Chevy-type restrictor kits won't work. Instead, the right-hand lifter bores must be individually bushed (super high-buck); or (lots cheaper) the right-side galley reamed for DC's press-in tubes, as described in the accompanying sidebar. In either case, an Allen set screw suffices for the left-hand galley.

While the production oil pump is adequate for street use, for all-out racing it must be upgraded to a high-volume/high-pressure configuration. Recently a new Chrysler high-output pump assembly was introduced that offers 25 percent more capacity than any previous A-motor, high-performance pump. This pump's capacity can be increased an additional 10 percent through the use of a modified B-engine shaft and rotor. To install the B-rotor and shaft, a special "Hi-Po" oil pump kit is required. Discard the rotor that comes in the kit and replace it with the B-rotor, machined down .050-inch; then use the rest of the kit to install the B-rotor and shaft into the high-volume pump's body (part Nos. in Parts List).

The trick pump setup is used with a standard Milodon or Moroso deep race

pan, which comes complete with the necessary extended pickup.

Cylinder Head

Chrysler W2 heads are recommended by Rossi, as they are worth 75 hp and .250-second over the best old-style high-performance heads, due to their improved intake and exhaust port de-

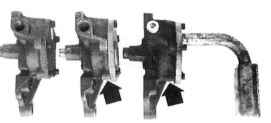

Compared to stock production oil pump (left), high-volume oil pumps can be identified by their thicker bodies (arrows). While normal pickup screws into hole in pump side, some aftermarket race pumps require a special cover with hole that accepts a centralized pickup.

As described in the text, a modified big-block B-engine shaft and rotor (right) may be used to increase the capacity of the Chrysler high-volume oil pump even more.

A small pressed-in plug (3462871) separates the oil filter's inlet and outlet passages. Often removed when cleaning the block, it is easily overlooked on reassembly. Make sure the plug is pressed into the block hole shown until it bottoms solidly (at least 2.3-inch below surface).

THE W2 CONNECTION

sign. The "Econo" version of the W2 head is more affordable than the regular (Basic) W2 head, since its cast-in rocker shaft pedestals allow the use of many economical stock-type valvetrain components. While the Econo head is no longer offered by DC, many local Chrysler enthusiasts and performance outlets still have stocks on hand.

If an Econo W2 head cannot be located, the next best choice is the standard "Basic" W2 head. Since it doesn't have the Econo head's cast-in pedestals, additional mods are required to avoid the need for special costly valve gear (see below).

Either head should retain "normal"-size 2.02 intake/1.60 exhaust valves.

Recommended valve job specs and seat width are in the spec chart. Besides a gasket match, the only porting required is minor pocket work behind the valve seat, in which the floor is smoothly blended into the seat window. Seal the heads to the block using the new No. 1009 Fel-Pro "O-ring"-style head gasket; it alleviates the previous need to O-ring the block deck, and can handle up to 13.6:1 compression ratios.

Camshaft & Valvetrain

To reliably achieve 9-second performance levels, Rossi feels that a roller grind is the only way to go. Direct Connection's P4120975 was specifically designed for Super Stock 360 engines us-

ing single 4-barrel carburetion, and as such is practically tailor-made for Rossi's needs as well. Nominally ground on 106-degree lobe centers, Rossi installed it 4 degrees advanced (104-degree center) to enhance low-end performance with his automatic trans. Keeping the cam correctly in phase is Speed-Pro's new ½-inch-pitch tooth, silent timing chain that provides roller timing chain-type reliability without the latter's undesirable harmonics.

Such a high-lift cam requires hefty valvesprings. The P3412068 valvespring is up to the task, and only costs $4 apiece. When used with DC's new Trick Titanium retainer that accepts a .070-inch longer installed valvespring height, the special (and expensive) extra-long valves, offset rocker arm shafts, and bushed rocker arms needed to work with special extended-length full-race valvesprings are not required. That, in turn, eliminates the need for full-tilt W2 heads with their special rocker stands. Instead, you can use the cheaper Econo W2 heads along with the unbushed Econo W2 offset right and left-hand intake rocker arms; 273, 340 Trans Am, or Econo W2 exhaust rocker arms; and 273, 340-TA, or Econo W2 rocker arm shafts with bananna grooves for improved lubrication. Since some of these parts are junkyard items, when compared to the full-tilt W2 package, the cost is considerably cheaper.

But what if only the regular W2 heads without the cast-in pedestals are available? Simple: Procure the P4120102 steel separate rocker shaft supports that come with offset holes for mounting the special high-buck offset (long valve) rocker shafts. Rotate the supports and drill new holes on-center. Voila! You

ROLLER CAM OILING SYSTEM MODS

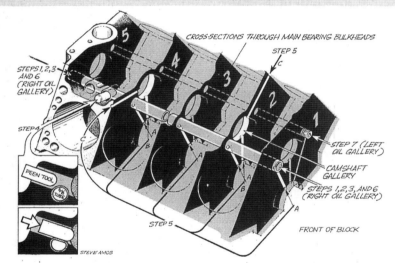

When roller cams are used in the A-motor, the oil feed to the tappets must be blocked while still maintaining the oil feed to the main and cam bearings. The cheapest method is to use the DC Drill and Ream Package (P4120602), and Tube and Peen Tool Package (P4120603). (Equivalent local-purchase substitutions are acceptable.) With the parts at hand, do the following:

1 Remove both front oil galley plugs and the right rear oil galley plug.

2 Starting at the front, drill the RIGHT galley ⅝-inch o.d. Go in approximately 10 inches, past four tappet bores. Repeat from other end.

3 Press the supplied ⅝ o.d. x ½ i.d. tubes into each end of the right-hand galley.

4 Drive the peen tool into each right-hand lifter bore to ensure adequate lifter clearance.

5 Using a .283-inch drill bit, open up the oil passages from: (A) right-hand oil galley to Nos. 1, 2, 3, and 4 main bearing bulkhead; (B) Nos. 2 and 4 main bearing bulkhead to Nos. 2 and 4 cam bearing bores; and (C) Nos. 2 and 4 cam bearing bores to block deck.

6 Tap for ½ pipe plugs on each end of the right-hand galley.

7 Drill and tap the left-hand galley front for a straight-thread Allen set screw. Loctite the screw in position at least ¼-inch past the oil feed from No. 1 bearing bulkhead.

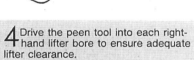

The 273-340 engines use a larger rear main seal than the 360. Since aftermarket race pans machined for the "weird" 360 are hard to come by, you can fabricate a low-buck spacer from Duro/Loctite "strip epoxy."

For his motor, Rossi chose the rear-sump Moroso Ultimate II pan. Note the extensive trap-door baffling and steering linkage pass-through hole.

Non-W2 heads—even the old 340 high-perf models—give away 75 hp to the new W2 castings. Here's one reason why: The two center exhaust ports are compromised to clear a head bolt boss (arrow), unlike the W2 heads (shown below).

can now mount stock-type rocker shafts on regular W2 heads.

Induction

True or false: To reliably run nines, you need a tunnel-ram and twin 4500 Holleys. Well, not in the A-motor's case. The Holley single 4-barrel W2 intake, along with the DC spacer plate to adopt this manifold to a Carter Thermoquad carb, is all that's required. No internal manifold mods other than port-matching is required. Rossi currently is running

an aftermarket 800-cfm Carter 8001, but any early '70s production TQ will do the job as well. For carb mods, see Critical Specs Chart.

Ignition

Rossi uses the DC aluminum-housing electronic distributor with tach-drive, and an oil pump driveshaft with a hardened tip and aluminum-bronze gear for use with roller camshafts. The brainbox is an MSD-7AL; MSD also supplied their new crossfire-eliminating suppres-

As illustrated by these gaskets, W2 heads' intake ports (bottom) are also all-new, their unique oval configuration requiring a special intake manifold.

CHRYSLER

RIGHT LEFT

INTAKE   EXHAUST

AFTERMARKET

LEFT RIGHT

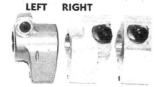

INTAKE   EXHAUST

Because of the W2 heads' revised push-rod locations, special offset intake rocker arms are required. However, old 273 mechanical or 340-TA rockers can still be used on the exhausts. Special Chrysler bushed rockers or Crane and Harlon Sharp roller rockers are also available, but they're simply not needed for this application.

Econo W2 head (A) is preferred for this engine because its cast-in rocker shaft supports permit the use of many less-costly stock valvetrain components. By contrast, regular W2 heads have supports machined off and use separate spacers that must be reworked to avoid costly (and unnecessary) investment in special high-buck valvetrain components.

From left: If using non-W2 heads on a tight budget, the '71 and later Carter Thermo-Quad cast-iron intake (2512099) works fairly well. Better for standard heads is the Edelbrock LD-340, if you can find one. W2 heads require a special intake made by Holley.

sion wires. Total advance with SoCal racing gas is between 36 to 38 degrees, checked at 3000 rpm. Rossi has found large plug gaps to be worth .03 to .04-second in the quarter. At sea level, plug gap is .055, but it may increase to as much as .100 at high-altitude locations like Denver.

Bulletproof high-output ignition system consists of DC aluminum distributor with magnetic pickup and tach-drive, controlled by an MSD7AL brainbox and associated components.

CRITICAL SPECS

(All dimensions in inches or fractions thereof unless otherwise noted)

MAIN BEARING	.002-.0025
ROD BEARING	.003-.0035
CRANK ENDPLAY	.002-.010
PISTON-TO-WALL skirt clearance, (forged pistons)	.008-.009
ROD SIDE CLEARANCE (aluminum rods)	.020-.025
PISTON-TO-TOP OF BLOCK (deck height)	.045 above deck
CYLINDER WALL FINISH	Sunnen AN 625
RING END GAP	
1st	.015
2nd	.007
GASKET COMPRESSED THICKNESS (Fel-Pro 1009)	.0385
PISTON-TO-VALVE (aluminum rods) (@0 lash w/check springs)	
Intake	.050
Exhaust	.075
VALVE JOB	
Seat width/intake	.050
Seat width/exhaust	.080
Angle/top	15°
Angle/seat	45°
Angle/bottom	70°
Combustion chamber volume	72cc
VALVESPRING PRESSURE	
@ installed height	140-145 lbs. @ 1.750
EXHAUST SYSTEM (automatic trans)	
Primaries	1¾ x 35
Collector	3½ x 18
CARBURETOR (Carter Thermo-Quad)	
Primary jet size	.098
Secondary jet size	.169
Air door spring	2½ turns
Float level	.900
Shooters	.037
Fuel pressure	6 lbs.
REDLINE	6800 rpm with 5.38:1 gears
CAMSHAFT (DC roller P4120975)	
Lift/intake	.630
Lift/exhaust	.650
Duration @ .050 tappet lift/intake	278°
Duration @ .050 tappet lift/exhaust	286°
Lobe displacement angle (lobe centerline)	106°
Valve lash	.038

Exhaust

Hooker has always fabricated Rossi's exhaust system, and his new Rampage is no exception. General dimensions for 360-based motors running automatic transmissions are given in the spec chart.

Transmission

When selecting engine components, attention must be given to the transmission type. Rossi has always preffered automatics due to their greater consistency. Currently, Paul uses the A904 Torqueflite, which he says is .15-second quicker than the traditional, beefier 727 Torqueflite. A "junkyard" 2.77:1 low gear for the A904 trans can be found installed in some late '79-'82 225-inch straight Six-equipped Aspens, Miradas, Cordobas, and Volares. For a converter, Rossi uses the 3800-rpm stall speed, 8-inch Turbo Action part carried by DC.

Performance Potential

Race engine building involves a lot more than merely bolting parts together. It requires careful, meticulous attention to detail. Particular care must be exercised in block preparation and component assembly. Using his years of engine-building experience and the previously described component parts, Rossi has achieved consistent 9.30's at 153 mph in his 2300-pound Wille Rells-built '84 Rampage. Now, admittedly we don't all have Rampages, but assuming a more typical 2800-pound Duster, and remembering that each additional 100 pounds will cost 1/10-second in the quarter, consistent 9.90's are well within reach with a chassis setup as de-

SOURCE: **Paul Rossi's Track-Ready Cars**
1826 N. Windes Dr.
Orange, CA 92669
(714) 633-4483

scribed in the DC racing manuals. And this is with a single 4-barrel Carter! The package looks even better when its total cost is considered: according to Rossi, if you did the assembly work yourself at home, it can be put together for under $4000! That's definitely Grade A performance. **HR**

MAJOR PARTS REQUIRED

(All parts and part numbers are Chrysler Direct Connection unless otherwise noted)

PART	SOURCE OR PART NO.
Used 360 block and crank	Junkyard
NASCAR .010-over forged pistons	
Cyls. 2-3-6-7	P4007856[1]
Cyls. 1-4-5-8	P4007857[1]
Aluminum connecting rods	Bill Miller
Big-block Chevy thin-wall piston pins	Venolia
Main bearings	
Standard	P4286956
.001 under	P4286951
.010 under	P4286959
Rod bearings	Sealed Power CB460M
Piston rings	Speed-Pro R9281+5
Drill and ream package (for oil galley mods)	P4120602
Tube and Peen tool package (for oil galley mods)	P4120603
High-volume oil pump	P4286589
B-engine oil pump rotor and shaft	P4007819
"Hi-Po" oil pump kit	P4120995
Ultimate 2 rear sump 7-qt. oil pan	Moroso 2107
W2 cylinder heads	
Econo (1st choice)	P4120043[2]
Standard "Basic" (2nd choice)	P3870812
Intake valves, 2.02 o.d. head	P3690230
Exhaust valves, 1.60 o.d. head	P3690231
Valve stem lock	P4120620
Head gasket	Fel-Pro 1009
Roller camshaft, 360 Super Stock	P4120975
Timing chain and gears	Speed-Pro 220-4102
Valvespring	P3412068
Trick Titanium titanium retainer	P4007178
Econo W2 rocker arms	
Intake, right-hand (4 required)	P3870825
Intake, left-hand (4 required)	P3870826
Exhaust (8 required)	P3870827
Rocker arm shaft	P3577086
W2 steel support package (for std. W2 heads)	P4120102
Pushrod kit, with assembly tool	P4007284
Roller lifters	Crane 66515-16
Holley Strip Dominator W2 intake manifold	P4007664
Carburetor (spread-bore Thermo-Quad)	Carter 8001
Carb-to-manifold adapter	P4007522
Tach-drive distributor	P4120701
Oil pump driveshaft (high-perf for roller cams)	P3690874
Electronic ignition control box	MSD-7AL No. 7210
Suppression wires	MSD 3118
Pro-Power coil	MSD 8201
Exhaust system (custom for chassis)	Hooker Headers[3]
Turbo-Action torque converter	P4120891
Oil	Quaker State 20W-50 racing

1. Discontinued, procure from Herb McCandless Performance Parts (Hwy. 54, Cherry Ln., Graham, NC 27253).

2. Discontinued, procure from Chrysler high-performance parts outlets.

3. See spec chart for dimensions.

OPERATION RUMBLE BEE

HEMI LIVES!

REBIRTH OF THE ELEPHANT

By David Freiburger

On December 6, 1963, a group of Chrysler engineers huddled to watch a pair of mammoth valve covers quiver as the 426 Hemi rumbled to life for the very first time. They had no idea that the lump-in-your-throat, tear-in-your-eye, fear-in-your-heart emotion of the day would be shared by car freaks every time a Hemi was fired up for the next 30-plus years. The engine would eventually shatter all records at its 1964 Daytona 500 debut, knock down the 200-, 250- and 300-mph barriers at the drags, run more than 400 mph on the salt and power the most violent street cars ever produced. Which is why Chrysler refused to make limp-wristed smog Hemis to meet '70s emissions laws. The last Hemi car was built in 1971, and Chrysler would later throw away the engine tooling. But the track legacy, street-race intimidation and limited production had secured the Hemi as an icon of power.

By the late '80s, Mopar Performance realized that the Hemi's legend lacked nothing but a future. Enthusiasts and profiteers were hoarding high-priced Hemi parts for chalk-mark-correct restorations rather than for going fast, making noise and sending Goodyear smoke signals. The bones had long since disappeared from the Elephant graveyard at Chrysler, but Mopar Performance made a commitment to re-create the Hemi. As a result, Mopar Performance now offers almost every part you need to assemble a brand-new Hemi!

An all-new 426 was assembled and fired in front of the crowd at the 1994 Mopar Nationals. And it was displayed in ads nationwide: "Hemi Returns. Power Is Restored." But restoration had little to do with it. The new Hemi was released to *compete* again with the Fords and Chevys. Readily available parts make it possible for Hemis to effectively compete in bracket racing, NHRA sportsman drag racing, or even Fastest Street Car racing. And that's our plan. This month

we'll take a look at the core parts that make up the new Mopar Performance Hemi. Next month we'll get together with Dick Landy and follow his buildup of a 750hp Street Hemi, then we'll show you how to install the engine in a '70 Super Bee and tune it for the strip. The Hemi lives!

CYLINDER HEADS

While symmetrical-port heads are tricky high-dollar race pieces for Chevy mills, they're O.E.M. equipment for Hemis. It's part of what makes them work so well and look so imposing. The new MP Hemi iron heads are nearly the same design as the original Street Hemi heads and look just like them except for the shape of the alternator-bolt boss. The new heads are stiffer than the originals for better head-to-deck sealing. The port shapes are similar and flow slightly better, and although the volume is a bit smaller than stock to pass NHRA tech, the walls are thicker to allow more porting. The heads come bare but are intended for the stock valve sizes of 2.250 and 1.940. The iron heads come standard (part No. P4529898), dual-plug (P5249524) or with extra-hard exhaust seats (P5249525).

We can also confirm the rumor about the new Hemi aluminum heads; they should be in warehouses by the time you read this. The original aluminum heads saved about 60 pounds over a pair of iron ones, and we expect the same today. We've heard that the lighter heads will have a slightly better exhaust port too.

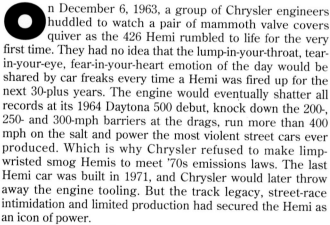

BLOCK IMPROVEMENTS

Since Chrysler chucked the original tooling, Mopar Performance was forced to redesign the Hemi when it was rereleased. As a result, the new blocks are the strongest iron Hemis ever. They feature .200 inch more deck thickness than the originals and revised head-bolt bosses to prevent the bolts from bottoming out or interfering with the water jackets. The cylinder walls are thicker—especially in the major thrust direction, making the block much more enthusiastic about ring-to-wall sealing. Original Hemi blocks with casting dates older than 1-19-70 have thicker main webs for additional bottom-end strength; the new block shares this design and further beefs it with filled-in webs (similar to the new A-engine race block). The iron used is the stiffer high-nickel iron used in original Race Hemis, and every new block is leak-tested (though not stress-relieved—an "X" cast into the side of an original Hemi block means it was stress-relieved by the factory). Hemi blocks are designed with extra material where a bolt hole should be for a Wedge block so that the Hemi may be modified to use wedge heads (although 426 wedge-specific two-bolt-main blocks are now available

too—under part No. P4529851).

The machining on the new blocks is of higher quality than the originals, but the specs are the same: 4.250-inch standard bore, 4.8-inch bore centers and 10.725-inch deck height. Freeze plugs come installed, and though we've seen some early units without cam bearings and main caps, those are supposed to be included as well. As with the originals, the Nos. two, three and four main caps are cross-bolted. In '66 and '67 some blocks came with cast-iron main caps, which are identified by the parting line on them. These caps should not be used in performance applications, so new ductile-iron main-cap sets (part No. P5249134) are also available.

When the new Hemi was first available, Mopar took a baseline dyno engine that was built from original parts, moved all the guts into a new block and made sure all the specs were the same. The new block made 25 more horsepower, which was attributed to the stiffer cylinder walls, deck surfaces and main webs. Additionally, Mike Landy at Dick Landy Industries tells us that the company has had new-generation Hemi blocks running 1000 horsepower with no problem. And for even nastier applications, a siamese-bore block will be available soon.

New Hemi block castings look almost the same as the originals, and unknowing cruise-scene droolers will never know the difference. An "M" has been added to the original 2468330 casting number, and three external ribs are missing from the right side of the block. Other stampings, nicks and parting lines are also noticeably different, but original Hemi parts work perfectly with the new block.

By using a new Hemi block with wedge-style motor-mount ears (part No. P4529852, shown), a Hemi can be bolted into a V8 B-body or E-body without an expensive Hemi K-member. A-body fans will have to swap K-members but can use an aftermarket big-block unit, which is vastly cheaper than a Hemi part. Mopar also sells the P4529850 iron Hemi block with original-type Hemi ears.

INTAKE MANIFOLDS

While the standard Street Hemi intake was a dual-Carter dual-plane arrangement, the early Race Hemi efforts demonstrated a lot of intake experimentation. One of the 1964 NASCAR manifolds was a single-four-barrel dual-plane. While not the same design as the NASCAR unit, the only Hemi intake currently offered by Mopar is an M1 single-quad dual-plane. The P4452034

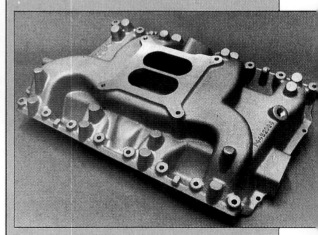

aluminum manifold makes carb tuning much easier than the original setup and is designed to make the same power when used with an 850 Holley. Driveability and low-speed torque are said to improve.

THE POWER COMBO

Mopar Performance (MP) is well-known for putting together proven factory power packages. And there's a rumor that a Hemi crate engine may be in the works. Playing off both those themes, we're going to follow along as Dick Landy Industries puts together a 750hp pump-gas Street Hemi for the '90s. The engine will also be used to test some new parts, since today's Hemi race-parts development is as exciting as that of the '60s. Next month we'll reveal what they are and follow the buildup in slobbering detail, but here's a parts list to whet your appetite.

QUANTITY	PART NO.	COMPANY	DESCRIPTION
SHORT-BLOCK			
1	P4529852	MP	Iron Hemi Block with Wedge-type motor mount
1	P5249503	MP	4.15-stroke full-radius Hemi crank
1	P4452776	MP	Fluidampr harmonic balancer
1	P5249557	MP	Heavy-duty 2.25-inch crank bolt and washer
8	P4529094	MP	Hemi connecting rod
1	P3412037	MP	Babbit main bearings, standard size
8	P2836184	MP	Babbit rod bearings, standard size
1	145-6001	ARP	Rod bolts
8	Special-order blank	JE Pistons	Custom pistons by DLI
1	R9745-65	Speed-Pro	Piston rings
1	P5249259	MP	Block hardware package
1	P4271961	MP	Rear-main seal
1	P1737725	MP	Distributor-shaft bushing
1	P4529404	MP	Chrome fuel-pump block-off plate
1	P4349816	MP	Chrome timing-chain cover
1	P4452795	MP	Timing cover bolts
1	P3412083	MP	Hemi gasket set

QUANTITY	PART NO.	COMPANY	DESCRIPTION
CYLINDER HEADS			
2	P5249525	MP	Iron Street Hemi heads with hard exhaust seats
1	P4529978	MP	Head bolts and studs
1	P4120247	MP	Head gaskets
1	P4529338	MP	Chrome Hemi valve covers with breathers
1	P4529897	MP	Valve cover studs
VALVETRAIN			
1	Custom	Crane Cams	.659-lift roller cam
1	66542-16	Crane Cams	Roller lifters
1	12559	Milodon	Torrington bearing
1	1600-TB	Isky	Thrust button
1	P5249269	MP	Timing chain
16	Custom	Smith Bros.	Custom-length pushrods
8	P4529705	MP	1.57:1 intake rockers
8	P4529706	MP	1.53:1 exhaust rockers
2	P5249631	MP	Intake rocker shaft
2	P5249632	MP	Exhaust rocker shaft
1	no part No.	DLI	Rocker-arm retainers
1	P5249505	MP	Rocker-shaft supports
16	P4120783	MP	Trick Titanium 10-degree retainers
1	P4529038	MP	10-degree keepers
1	P4120635	MP	Lash caps
16	P4007536	MP	"Chrysler triple" valvesprings
8	Custom	Manley	2.320 titanium intake valves
8	Custom	Manley	1.90 exhaust valves
OILING SYSTEM			
1	P3412026	MP	Milodon dual-line kit
1	P3690876	MP	Hardened oil-pump driveshaft for roller cam and Milodon pump
1	P4529190	MP	Race oil filter
1	Custom	Stef's	Fabricated aluminum oil pan and scraper
1	P4120613	MP	Oil pan bolts
INDUCTION			
1	Custom	DLI	Fabricated intake manifold
1	P4529977	MP	Intake bolt kit
1	P4529431	MP	Manifold heat shield
1	Custom	Brad Urban's Carburetor Shop	Holley Dominator
IGNITION			
1	P4120942	MP	Race electronic distributor
1	P4349279	MP	Chrome distributor hold-down
1	3128	MSD	Plug wires
1	8201	MSD	Pro Power Coil
1	7220	MSD	7AL ignition box
1	P4120294	MP	Hemi spark-plug tubes
COOLING			
1	P4286900	MP	Aluminum water-pump housing
1	P4529102	MP	Aluminum water pump
1	P4286759	MP	Water neck

CONNECTING RODS

Engineering papers from the '60s indicate that a stock Race Hemi at 7200 rpm places a 16,000-pound separation load, or tension, on the connecting rod. Therefore, the rods are quite stout from the factory, and the new Mopar P4529094 forgings are no exception. The rods have the original 6.86-inch center-to-center length and high-strength steel 7/16 bolts. The rods will not work for 440s or 383s, which use 6.76- and 6.36-inch rods, respectively. The old 7.06-inch NASCAR Hemi rods with 1/2-inch bolts are not available.

CRANKSHAFTS

There were three types of production Hemi cranks. The most common is the '66–'71 Street Hemi version with undercut fillets (where the journals merge into the balance weights). The second type is the famed Kellogg crank with full-radius fillets for strength. These were only Tufftrided .003-inch deep, and early manuals do not recommend having them machined. The final is the NASCAR crank that's similar to the Kellogg unit but with a different snout. Since each of these cranks is either rare or undesirable for all-out performance, Mopar now offers a full line of brand-new Hemi cranks to choose from:

P5249206   3.75 stroke, undercut radius (resto)
P5249207   3.75 stroke, full radius, balanced
P5249208   4.15 stroke, full radius, balanced
P5249503   4.15 stroke, full radius, unbalanced
P4529095   3.75 stroke, full radius, replacement
            for P3690285 Kellogg crank, unbalanced
P5249164   Unmachined 1053 steel forging

Each of these eight-bolt cranks is made of 1053 steel, and they're now nitrided rather than Tufftrided. The Hemi cranks have the same main- and rod-journal diameters as a 440 and can be used in RB wedges. It's becoming popular to make stroker low-deck B-engine (383 or 400) wedges, so Mopar also has stroker cranks to match the smaller 2.62 main-journal diameter of the B engines. The selection of cranks and pistons makes 500ci Mopars more practical than ever. MP also sells O.E.M.-type balancers (part No. P3830183) and Fluidampr SFI-approved dampers (P4452776), as well as a heavy-duty 2.25-inch crank bolt and washer (P5249557).

PISTONS

Since light pistons make power and original Hemi pistons are rather heavy (at about 843 grams), the most recent development at Mopar has been to lighten the Hemi pistons without sacrificing strength. Now there's a full line of pistons designed to be used with the compression ratios found in original Race Hemis and Street Hemis, as well as new 9.0:1 units for use with pump gas. There are also pistons with altered pin heights for use with the MP stroker cranks and stock rods. Some of the new pistons do not have an offset pin, so one part number will work on either bank.

PART NO.	INTENDED COMPRESSION RATIO	BORE	STROKE	GRAMS
P5249566	12.5:1	.060	4.15	758
P5249567	9.0:1	standard	3.75	807
P5249568	9.0:1	standard	4.15	774
P4529107-8	10.25:1	.060	3.75	856
P5249436-7	10.25:1	standard	3.75	780
P5249564-5	12.5:1	.060	3.75	787

FLYING ELEPHANTS

Designing the 426 Hemi started 13 months prior to the 1964 Daytona 500, and Richard Petty's record-breaking winner used an engine block that was cast just 13 days before the race. Despite hurried beginnings, competitors and NASCAR spent the next several years figuring out how to stifle the Hemi. The 1965 Hemi ban by NASCAR and USAC is well-known, but the winged cars were also dealt a blow. With the then-new Holley Dominator carb, a Hemi and a swoopy Charger Daytona, Bobby Isaac set an all-time record of 201.10 mph at Talladega in 1970. NASCAR subsequently dictated that aero cars would have to run 305ci engines instead of the 430ci engines for traditional cars. Rather than detune the Hemi, Isaac headed for the salt. The same Daytona ran 217.37 at Bonneville and set 28 national and world records.

VALVETRAIN

Every Hemi valvetrain part is available, including six different camshafts and the related matching components. This latest additions are rocker-shaft sets (part No. P5249631 intake, P5249632 exhaust) and the P5249633 rocker-shaft plug kit,

so you can clean out the shafts while rebuilding the engine. The Hemi requires two of these plug kits. **HR**

SOURCES
Dick Landy Industries
Dept. HR12
19743 Bahama St.
Northridge, CA 91324
818/341-4143
Mopar Performance
Dept. HR12
P.O. Box 215020
Rochester Hills, MI 48321
810/853-7290

By Kevin Boales

HOW TO BUILD A ~~STREET~~ STOCK ELIMINATOR

Bob Lambeck's A-Engined Low-Dollar Screamer

So, you wanna go racing? Don't have much green? All you have is a grocery-runner and moths in the wallet? Well, you've come to the right place, buddy!

Some of you may have heard of a guy named Bob Lambeck, a 20-year veteran of the asphalt jungle who's run MoPars for Dick Landy, raced in about every full-bodied class ever devised, and is now successfully building engines for boats and race cars (Bob Lambeck Enterprises, 7510 Gloria St., Van Nuys, CA 91406, 213/787-5678). Having started long ago with a low-buck '57 Chevy stocker, Bob can relate to the dilemma a lot of us face when we get the urge to go racing. He's decided to re-enter the drag racing arena with a new project, and it's going to appear step-by-step here for *you'se to peruse*.

What we're going to show you is how to build a potential class-winner that runs in the high 11s, can be driven between race days as a street machine, and shouldn't keep you in the financial dumper for long. The entire project will cost about six or seven grand, depending on how you come across your parts and how carried away you get with the details. Bob likes to do this kind of "grass roots" race car because the thing can be pounded down Main Street all week and raced on Saturdays or grudge nights. It'll run real well through the muffs, too.

The basic equipment? A well-used '74

Duster with a 360 and an automatic. The problem? How to build a car within the tech requirements of the NHRA H/Stock Automatic Class. No cheating allowed; we're gonna stay completely legal eagle on this one.

To get hold of the NHRA tech sheets and "blueprint" requirements, contact the NHRA, 10639 Riverside Dr., North Hollywood, CA 91602, and tell them the make, year, displacement, and carburetion on your car. They'll charge you a couple of bucks for the tech sheet and another two for the "blueprint" specs (they won't actually send you a blueprint;

you'll receive a list of measurements the tech inspectors use during inspection to verify certain things about your car). Also, you should have a copy of the current NHRA rulebook as a guide to preparing the car (available from the NHRA for $4). Now, on with the story . . . .

Engine Prep

To be legal, the engine must remain essentially stock. That means our project car must have the same innards as supplied when the car was new, with a few important exceptions. Bob "blueprinted" the block, cleaning up the cylinders with a 0.030-inch overbore, creeping up on the finished size with the Sunnen CK (Cylinder King for those of you who like to decipher initials) -10 hone. Bob feels by far the most important aspect of any block prep is to assure good ring seal; he does

Photography: Kevin Boales

The cave. The Duster's engine compartment was not pretty to begin with, but in an upcoming installment you'll see an entirely changed place for the engine to live.

Bone stock—that's what this story is all about. The '74 Duster looks kind of plain, doesn't it? Would you believe it'll run in the 12s a month from now?

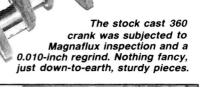

The stock cast 360 crank was subjected to Magnaflux inspection and a 0.010-inch regrind. Nothing fancy, just down-to-earth, sturdy pieces.

The finished block, ready for assembly. Bob paints the valley to seal the iron and speed oil flow back to the lower end. Looks nice, too.

Humble as it is, we used stock Chrysler pistons (part No. 04026453) and pins. Although the NHRA says the ID number must be visible, these pistons have no factory stamped number. The inked indentification will go away before the end of the first pass.

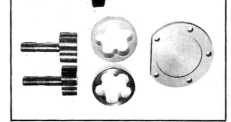

This photo shows the difference between the stock oil pump and the DC part No. P4120995 high-performance oil pump assembly. This type of aftermarket pump is considered OEM design and is NHRA legal.

Bob spends a lot of time laundering the block after all the machine work is complete, using good ol' Tide and then lacquer thinner to degrease the casting and remove all the remaining grit from the hone.

that by running about 25 degrees of crosshatch in the finish hone, which coincidentally is the angle the CK-10 produces when set to Sunnen's specs.

To really do the job right, Bob says you've got to get all the remaining debris out of the hone pattern with a thorough washing—first with our old friend Tide and then with lacquer thinner. You'll be amazed at the amount of stuff left over in the cylinders otherwise. The trick to good ring sealing is to keep the rings from contacting the cylinder wall; they should be riding on a film of oil instead. By cleaning out the crosshatch, the oil has a place to live and the rings will seal perfectly.

Bob align bores the block, squares it up by making the decks parallel to the

This is about the only place in the engine assembly where using a die grinder is okay. Bob matched the passages between the rear main cap and oil pump outlet with a little grinding. Stock pickup is used because we couldn't change the oil pan under NHRA's rules.

Both Moroso and Milodon make lifter valley oil baffles like this one. The idea is to keep the hot oil from spraying all over the underside of the intake manifold.

STREET ~~STOCK~~ STOCK ELIMINATOR

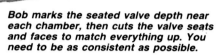

Hooker made the headers for the car, including the weird setup for the left side of the engine, designed to clear the steering gear. We'll be using a collector extension to make a total collector length of 24 inches.

Bob marks the seated valve depth near each chamber, then cuts the valve seats and faces to match everything up. You need to be as consistent as possible.

Although the pan must be stock, NHRA rules allow you to install baffles to keep the oil under control. Bob fabricated these.

The only item in the picture here that isn't Chrysler is the lifter—it's a Crane juicer for the 360. The cam in Bob's car is a Crane item, too, but you won't find it in the catalog because they ground it for this engine. In Stock Class, the tech inspectors will check for duration, lift, and valve opening/closing points. The rocker arms are DC parts, P3690711 (left hand) and P3690710 (right hand), and legal. Tulip-shaped valves are stock 360 items.

Direct Connection stuff is everywhere on the car, including the Chrysler Sure Grip differential and 4.80:1 gear set. This is an 8¼-inch set (DC part No. P3690441).

crankshaft bores (the cylinders are cut perpendicular to the crank centerline, also), and sets the deck height according to the tech sheet. The deck height is determined somewhat by the thickness of the head gasket you'll be using; thin gaskets must be compensated for by an increase in the deck and vice versa.

The lubrication system receives some attention at this time, when Bob removes all the oil gallery plugs to gain access to the passages. He has a set of small brushes to scrub out the passages; then he re-plugs them with threaded plugs. Water jacket plugs are replaced with copper models rather than steel; Bob just likes them better. The oil pump pickup can be chamfered to match the passage in the bottom of the block (see photo). The finishing touch is to paint the inside of the block with Rustoleum Damp Proof Primer or Ditzler DP-40 primer to seal off the pores in the iron. Keep the paint away from finished surfaces, including the lifter bores, cam bearing bores, etc.

Install the cam bearings and check the fit of the cam; it should rotate easily in the new bearings. If it won't, look for "burnished" areas on the bearing surface and smooth them out with a 3M Scotch-Brite green pad. With that taken care of, let's

turn our attention to the heads.

You can't do much to the heads but blueprint them; the main thing is to keep your die grinder in the next room while you build your stock engine. You may replace the valves with aftermarket items if they're the same size and shape as the OEM valves, and they can be ground in the usual three-angle way, but you can't use valves with reduced stem diameters or tulip heads unless the engine came from Motor City that way. Absolutely no rework can be performed on the ports; in fact, you can't even glass bead the inner surfaces of the head for clean up. Bob goes through some interesting steps to equalize the chamber volumes in each head. The valve seats are cut to exactly the same depth through the use of an old valve (one for the intake, another for the exhaust) with a ball bearing welded on the chamber side of the head, which he places in the seat and measures from the ball to the head surface. He then cuts each seat to get the same depth from one cylinder to the next. Each valve is cut so the distance from the chamber face to the seat contact area is the same for each one. These two steps will assure similar

chamber volumes by locating each valve in the same relative position. Again, no cleanup is allowed in the chamber itself, so be careful about keeping each chamber on an equal basis with the next.

You can use aftermarket valvesprings, retainers, and locks, and you can replace the rocker arm studs with threaded types if yours aren't already that way. Rocker arms must be stock ratio (no roller allowed), but can be aftermarket forgings. Pushrods cannot be aftermarket types; they must be OEM. The tech inspectors will check two things on your valvesprings: seat pressure (the pressure with the valve closed) and the pressure at a specified height, which varies from engine to engine. If your engine was equipped with double springs or dampeners when new, you'll be allowed slightly more pressure at the second check height.

The camshaft will be checked, and here's where the men are separated from the boys. They'll check to see if the cam is within the specified lift, and if it's not within the allowed amount, you'll go down the drain. The tolerance is 0.002 inch. Don't cheat. The lift spec appears on the NHRA tech sheet.

Bob likes to run a baffle in the lifter galley to keep hot oil from spraying all over the bottom of the intake manifold; the rules will allow you to do the same if you'd like. The intake manifold must be the same as stock, including the material. You can't run an aluminum intake unless the car came with one in the first place. Don't bother matching the ports on the manifold to the gasket; you'll earn yourself a long ride home for it. The carbure-

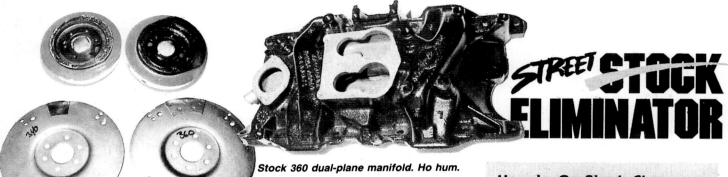

Stock 360 dual-plane manifold. Ho hum.

Bob likes to keep the balance weights off the torque converter and on the flexplate. A-1 plates show differences between 340 and 360 engines, with larger balance weights for the 360. Balancer also has a much larger weight cast into it for the larger engine.

MSD Cap Adapt fits with a little file work on the DC part No. 3690200 distributor. Bob's going to use MSD's coil, 7AL ignition, and an Autotronics/MSD retard device to back the timing off after the car gets into the top end. The thinking behind the larger rotor and cap is to move the terminals inside farther apart in an effort to avoid any spark leakage to adjacent terminals. Compare it with the stock cap and rotor.

tor has also got to be the same as stock. It can be a replacement, but it better be the same model and manufacturer as the OEM job. The carb may be modified for drag racing.

You can run any oil pan baffle you want, as long as it's in a stock pan. Windage trays are okay, too. You may run aftermarket headers (Hooker provided the split wonders for Bob's car). Mufflers are up to you.

Body Prep

Basically the body must remain (you guessed it) stock. Wipers and visors are optional, but all the accessories must still remain in the car, including a heater for cars built after 1968. One interesting point is that if you want a rollbar, you can take out the rear seat to make up the weight difference. The exposed area under the former seat *must* be carpeted though. Obviously, the NHRA is interested in preserving the beauty of drag racing. No fiberglass bumpers, no Lexan windshield, no tin interiors.

Chassis Prep

In the interest of maintaining all that was once show-room fresh, the NHRA won't allow you to do much to the chassis, but you can do *something*. Bob figured he didn't like the idea of going 130 mph in a car with 80,000 grueling miles on the dash, so he decided to go through all the suspension rubber, replacing anything that moved—all the front end bushings, ball joints, and tie rod ends. The rear springs were replaced with new versions, including the bushings. You can legally run bolt-on traction bars, although the rules won't allow ladder bars or more sophisticated rearend systems. You must have a driveshaft loop to avoid the dreaded "eighth-mile pole vault" exercise. If you're really into suspensions, don't bother getting into them on this car unless you want to be disappointed. The location points for the front suspension must remain (what's the word again? . . .) stock. The same goes for the rear springs.

The rear axle must be the same as the one born in the car, but you can run any factory optional gear ratio. You can also run posi whether or not it came on the car. Bob found a couple of used axles in the local "salvage yard" (they're projecting a new image these days), yielding all the components we needed for the Duster. He also managed to dig up a small ring-and-pinion set from one of his old cars with a 4.80:1 ratio. You can run aftermarket axles if you want, though we didn't want to.

NHRA's thinking behind the Stock Class is to equalize the little guy's chances against the full-time racers. By limiting the number of modifications possible, they have effectively eliminated any chance of someone dominating the Stock Classes for very long. It's a great form of racing—relatively cheap, fair, and fast enough to let you enjoy it.

In our next installment, we'll show you how the car was finished, taken to the track for dial-in, and finally run at the NHRA Winternationals in February. We'll also go into how to get through tech inspection, how to keep good records, and how to speed up your troubleshooting in the pits. **HR**

How to Go Stock Class Racing—Do's and Don'ts

You Can	You Can't
Remove the air cleaner	Use Ram tubes or fabricated ducting
Blueprint the engine	Stray too far from dead stock
Use an aftermarket cam	Exceed the lift and duration specs in the NHRA Tech Bulletin
Use an offset key to correct cam timing	Use a gear drive for the cam
Use aftermarket lifters	Use other than OEM-supplied type (hydraulic or solid)
Replace the carburetor	Use a different model, brand, or year than the OEM carb supplied
Resize the air and fuel passages in the carb	Resize venturis
Punch the cylinders out a maximum of 0.040 inch	Lighten any parts except for normal balancing
Stroke the crank 0.015 inch	Exceed NHRA-specified chamber volume
Run uncorked	Use larger than a 4-inch collector
Use an aftermarket fan if you put it in front of the radiator	Remove the belts, resize any pulleys, or loosen the belts excessively
Use an electric fuel pump	Remove the stock mechanical pump
Use a fuel pressure regulator	Mount a cool can within 6 inches of the firewall
Use a carb insulator-type of base gasket	Use more than two base gaskets or non-OEM spacers
Run any pump gas; it'll be subject to check by the tech inspector	Replace the fuel tank or move it from its stock location
Add baffles to the oil pan	Increase the oil pan capacity from stock
Use any OEM or OEM aftermarket oil pump	Run a dry sump oil system
Run either cast or forged pistons	Run anything lighter than the stock piston assembly weight
Use a torque strap (one)	Use an engine plate
Use any valvesprings you want as long as they appear stock	Exceed the NHRA-specified spring pressures for your engine
Run any automotive plugs and wires	Replace the radiator with a lighter unit/OEM only
Use any throttle or choke linkage as long as it's positive action	Change the intake manifold

MAGNUM MIRADA

By Marlan Davis

Paul Rossi has Chrysler 440-building down to a science. Follow his recipe, and yours will really cook. Phase I engine (left) is suitable for street/strip competition, uses

Edelbrock Street Tunnel Ram. Super Stock engine (center) likes Weiand Six-Pak Tunnel Ram—hood clearance permitting. Ultimate intake is fully adjustable Mullen 8-bbl.

MoPar's big-block B-motors have a lot going for them. For one thing, they're built by Chrysler. That means they're rock solid, beefy and dependable. It also means they're easily obtainable and can usually be picked up dirt cheap (usually for under $100 complete) in thousands of wrecking yards all across the country. Paul Rossi (whose trick Super Gas Mirada chassis we covered last month) has cooked up more of these motors than just about anyone. He has developed a "recipe" that should enable any MoPar racer to be successful in ET brackets, Super Stock or Super Gas racing. Now he shares all his speed secrets with our readers.

If that's not enough, Paul says he's willing to sell completely assembled "cookbook" motors for under $4000 to all comers, less intake and carburetion. All you have to do is send the block, crank and heads to him at the Carburetor-Electric Service Center (3006 Harbor Blvd., Costa Mesa, CA 92626, 714/546-3265). If you live back East, the same engines will also be available from Dave Koffel (9165 Longcroft, Union Lake, MI 48085, 313/363-5239). Finally, for those who want to do all the work themselves, Bill Bagshaw's ProParts

(4113 Redwood Ave., Los Angeles, CA 90066, 800/421-9931) will have all the parts mentioned in this article under one roof.

Phase I: Dual-Purpose Street/Bracket Car

Before doing anything, get a copy of MoPar's newly revised "bible," the *Direct Connection Performance Book* (part No. P4120792). It tells everything that's needed to build Chrysler products for high-performance use. Don't let your MoPar leave home without it!

MoPar big-blocks have two different

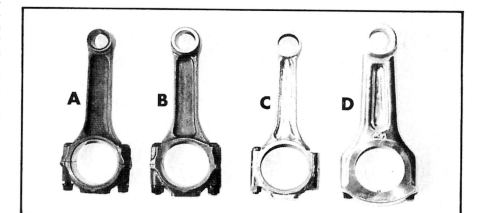

Rod choice depends on many factors, including sanctioning body rules and piston weight (the lighter the piston, the lighter the rod). P2531589 Phase I forging comes Magged and double-shotpeened with 7/16-inch high-strength nuts and bolts (A). Phase II uses improved 440 Six-Pak forging (P3690649) that also comes Magged and shotpeened, but with small end bushed for floating pins and flash ground off beam ends (B); it accepts ⅜ bolts. Phase III steel rod (C) can be fully reworked 440 standard production forging (2406770) that's lighter than other factory rods, but cheaper aluminum rod (D) is preferred. Made by Bill Miller Engineering, it uses lighter .927-diameter small-block Chevy wrist pin to save even more weight. (continued overleaf)

MAGNUM MIRADA

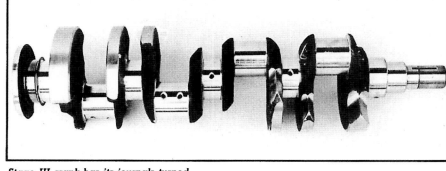

deck heights: 361/383/ 400 blocks measure 9.98 inches from the crank centerline to block deck, while 413/426/440 engines are 10.725 inches high. Most of Rossi's current Phase I and II motors are based on the Direct Connection (DC) 440 bare block, P4006628, although a junkyard block is suitable if it passes Magnaflux inspection. Rossi says he cures new blocks by letting them "sit outside on the patio for a year." All blocks are sonic checked to ensure they don't contain any air pockets or serious core shift. A 10 percent muriatic acid solution is used to remove any accumulated rust, then Ed Pink Racing Engines bores Phase One en-

Stage III crank has its journals turned down to fit in lighter 400 block. Smaller journals also reduce bearing speed and lessen friction. Crank is knife-edged by Velasco to lighten it and further reduce drag.

DC pistons (A) are most economical way to go racing. They come either standard (P3690833), .030-inch over (P3690834) or .060-over (P3690835). Use with ring set P3690922 (standard), P3690923 (.030-over) or P3690924 (.060-over). Similar pistons and rings are offered for 383 and 400 engines. Phase II (B) and III (C) slugs are ultra-light two-ring Venolia forgings. Phase III pistons must be ⅝ shorter to fit in low-deck block and are not recommended for street use due to possible piston rock at TDC. While Phase II slug is Super Stock-legal 10.5:1 flat-top, Phase III has slight .080-inch-high dome to help get compression up in 13:1 range.

gines .060-over. He also squares the deck and mills the oil pump mounting surface flat to ensure against any leaks. If necessary, the block is align-honed.

Paul thoroughly deburrs "every inch" of the block using a hand grinder. It often takes three to four days to polish the block's innards, but Paul feels the payoff (in terms of faster oil return to the sump and the avoidance of broken-off casting slag fouling up the oil system) is well worth the effort. After the polishing, all non-machined surfaces are given a coat of Rustoleum. One additional block mod is the use of special grade-9 Specialty Fasteners main bearing bolts. Half an inch longer than the stock bolts, they provide superior thread engagement and strength over the replaced stock items, and make cap studs or 4-bolt caps unnecessary.

A forged crank out of a '73-or-earlier 440 is preferred over the later cast cranks for serious performance use. The 440 crank based on the Hemi forg-

ing (P3690285) is the most durable crank of all, but its added weight compared to a regular forging is a definite drawback in drag racing. All cranks should be internally balanced, but don't go overboard with the Mallory heavy metal. Says Paul, "Up to 7000 rpm, I'd rather run the engine unbalanced than put the weight back in the crank. Unbalanced engines won't show up on the e.t. slip, but a couple of pounds of metal will." To save even more weight, cut down the counterweights. If you have two "identical" cranks, weigh each and use the lighter one. The cranks should be checked for straightness and Magnafluxed. There are no bearing problems with the B engine, so the crank doesn't seem to mind being ground up to .030-inch undersize as long as the oil holes are rechamfered. Turning the crank undersize also permits the rod journals to be indexed 90 degrees apart for the correct stroke. An undersize crank may also have less frictional drag. On the other hand, unit

Two critical porting areas on 440 heads: Intake ports are enlarged to their optimal cross-sectional area (A). Sharp edge in valve pocket area must be smoothed out (B). DC's home porting kit (P4120437) allows the careful do-it-yourselfer to essentially duplicate these results.

loads can be increased, but this has not caused any problems in Rossi's drag motors.

Clevite 77 Tri-metal bearings (DC P2836118, mains; TRW CB527P, rods) are preferred for Phase I applications. The mains are run tighter than the rod bearings (see spec chart) because oil to the rods has to flow through the mains first. The main bearings come fully grooved, so there's no need to cross-drill the crank. To minimize any chance of spinning a rod bearing, Paul drills a hole through the bearings to accept a special dowel pin installed in the connecting rod.

Rossi uses the DC P2532795 thin harmonic balancer, since it weighs less than other dampers. Both steel and alu-

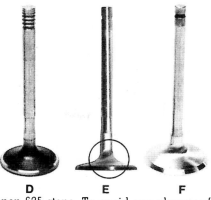

A B C D E F

Intake valves: Stock 2-groove 2.08 intake (A) can be replaced with DC P3690708 with single lock grooves so Mr. Gasket 10-degree valve lock can be used. Better performance will result with P4007942 2.14 intake (B). All-out engines need Mullen 2.19 intake (C), that comes with 5/16 stem for lighter weight and less flow restriction. Exhaust valves: All 1.74 standard-size valves (D) use 3-groove keepers. DC P4120579 (E) offers improved flow, thanks to 1.81 head diameter and thicker .100 margin (the transition from stem to head). Ultimate is 1.81 head, 5/16 stem Mullen (F).

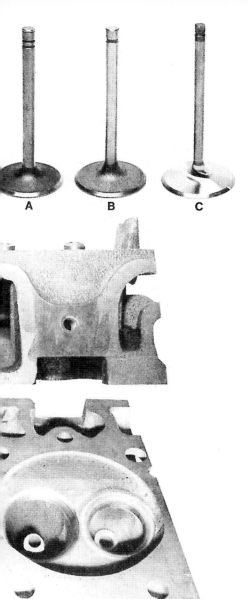

nen 625 stone. To avoid any chance of damaging the rings as they are installed, the sharp edges on the top of each cylinder bore are chamfered with a special Sunnen tool.

Two excellent hydraulic cams are available for Phase I engines. The Cam Dynamics 7000 Plus/No. 980 (DC No. 4007277) features 352/380 degrees intake/exhaust duration, .467/.483-inch lift and 150 degrees overlap. It was originally designed as a "cheater" cam for Stock Eliminator racing, and really pulls strong in a car with steep gears. The cam is also highly effective on the street; use it with hydraulic tappets (P4006767). DC's P4120237, with 292 degrees duration, .509-inch lift and 76 degrees overlap, makes a good bracket racer with a competition-type torque converter, and is furnished complete with hydraulic tappets. To drive the cam, DC offers excellent Magnafluxed roller timing chain and sprocket sets. P4120263 fits cams with three sprocket-attaching bolts, including the stock

(continued overleaf)

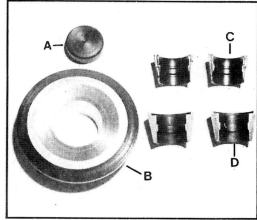

Mr. Gasket lash cap package (6164, 5/16 stem; 6166, ⅜ stem) (A) is needed with high-lift solid and roller cams. Stage I and II springs use Trick Titanium retainer (B), carried by DC. On those valves using single-groove keepers, stock 7-degree valve locks (C) ought to be replaced by Mr. Gasket 10-degree locks (D). Part No. 6160 fits 5/16 stems while 6162 is for ⅜ stems.

minum flywheels are offered for manual trans cars; a special flexplate (P2466326) is available to mate with the recommended Turbo-Action 3800 (P4007290) or B&M "J" automatic trans torque converters.

The choice in rods is similar to choosing a crank: Once again, the key is to get weight down without sacrificing necessary strength. For Phase I, the beefy P2531589 rod made from the Hemi forging is recommended when using the relatively heavy DC pistons. The rod's 7/16-inch nuts and bolts can be replaced with SPS parts (P4120070) for even greater strength. Offering an 11.5:1 nominal compression ratio, the DC slugs are machined from TRW forgings. Their ring grooves accept a Chrysler 1/16-1/16-3/16 set that features a moly top ring, cast-iron second ring and low-tension oil ring. The top two rings come .005-inch oversize to allow them to be trimmed and select-fitted to each bore. Wall finish with these rings should be obtained using a Sun-

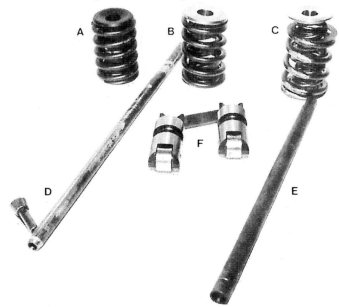

Valve springs should complement cam and not go into coil bind at max lift. Good through .620 lift is P2806077 Purple Shaft spring (A), while Bryant Racing Enterprises spring (B) works through .750. Crane Triple "877" spring (C) used in Phase III requires its own matching titanium retainer. For general high-performance use, DC's P4007284 "build-it-yourself" pushrod kit (D)

comes with 16 hardened ⅜-inch-diameter pushrods that can be cut to any length. A fixture is provided to hold pushrod while pressing in tip. Phase II and III use 7/16 chrome-moly pushrod (E) made on special-order basis by Crane or ProParts. Crane 66515 roller lifter (F) is superior because its welded guidebar keeps lifter from exiting bore in event of pushrod failure.

"cheater" mentioned above. "One-bolt" cams like P4120237 must use the P4120264 roller.

DC ⅜-inch-diameter "cut-to-length" pushrods will get the job done with these bumpsticks. Valve springs for Phase I are the factory P2806077 chrome vanadium Purple Shaft pieces, effective up to .620 lift. Rossi doesn't like the stock non-adjustable stamped steel rockers, since he feels they wear excessively. Replace them with Chrysler Max Wedge adjustable nodular iron rockers (2463242, right; 2463243, left) and rocker shaft spacer springs (2202557). Crane offers a kit that includes all 16 rockers and the necessary springs under part No. 64770.

These valvetrain parts should be installed on either the '67 small-chamber 440 heads (part No. 2806762, cast 2780915) or the far more common '68-'70 383/440 heads (part No. 2843904, cast 2843906). Both are about equal, since any compression loss in the latter head can be made up for by milling the block. Any '71-and-later heads are generally designed for low emissions rather than efficiency and should be avoided. Additional performance can be achieved by installing 2.14 intake (P4007942) and 1.81 exhaust (P4120579) valves, although the seats must be reworked. All Rossi's heads are ported by Mullen & Co. (340-C E. Carson St., Carson, CA 90745, 213/835-0686). The chambers are fully cc'd and the intake ports enlarged about 30 percent over stock. On the exhaust side, the port is cleaned up, but *not* enlarged. Mullen says, "The port is short, and an increase in area would be unstable," tending to create more turbulence. Do not match the port with the exhaust headers either, since that would effectively increase the port's area. A key area to watch on any 440 head is the valve pocket. The sharp edge left by the factory significantly impedes mixture velocity and should be smoothed out.

Normally, Mullen uses a three-angle valve job. With low-lift cams under .480-inch lift and the 2.14 intake valve, the angles range from 30 degrees on the seat to 60 degrees in the throat area. Seat widths are usually .060-inch on the intake and .050 for the exhausts. Bronze wall guides are installed in all Rossi's motors. With these guides, Rossi feels valve seals aren't required. The heads go on the block using Chrysler 3614284 beaded steel head gaskets sealed with Permatex Copper Coat and Specialty Fasteners' 12-point bolts.

As for induction, the Edelbrock STR-14 with Six-Pak top works best

Since combustion chamber and cylinder don't always line up perfectly, Rossi orders Venolia pistons with only roughed-in valve reliefs. He locates proper pocket location using headless "valve" with stem ground down to a point. Paul installs "valve" through guide and taps it sharply with a hammer at 10 degrees BTDC (for exhausts) and 10 degrees ATDC (for intakes), thereby finding exact center of each pocket.

Easy way to check piston-to-valve clearance: Required is a degree wheel (like this one from Mills Specialty Products), dial indicator and "checking springs" (available from most cam grinders). With rockers "zero lashed," piston is moved to its closest point near valve, then rocker is pushed down by hand to contact piston top. Distance dial indicator moved is the piston-to-valve clearance. Note just-introduced Crane roller rocker (it fits on Sharp's or Bryant's shafts) and Henry's Competition aluminum shaft support bracket that fits on pedestal without machining.

For serious competition use, replace stock rocker arm setup (A) with either Harlon Sharp (B) or Bryant (C) aluminum roller rockers and chrome-moly rocker shafts. Stock shaft supports are replaced by modified A-engine P4120102 steel rocker arm support package (D). Cecil Yother solid aluminum spacers (E) are used in place of stock rocker shaft spacer springs.

with Stage I engines. No matter what manifold is chosen, it should be matched to the heads' intake ports.

Rossi believes that street/strip cars need headers with 2-inch-OD x 35-inch-long tubes that dump into 18-inch-long, 3½-inch-OD collectors. Hooker adjustable race headers for B and E bodies fit these specs; No. 5319 fits '62-'65, while 5302 covers all '67-'72s. Both sets have variable 32-40-inch primaries and a 12-inch collector, with an 18-inch slip-on extension.

To fire the mixture, use either the cheaper P3690201 Chrysler pointless

electronic tach-drive distributor with mechanical advance only, coupled to the Autotronics MSD-7C Multispark ignition, or the lighter Accel BEI-3 distributor with the MSD-7A. Accel 9mm Fat Stuff II wire is needed to contain the hot spark. With this setup, plug gap can be increased to .080, resulting in improved elapsed times.

All these pieces need an effective oiling system. B engines don't require thick high-viscosity oil or trick additives; 30W Valvoline racing oil will do the job. The stock pan is adequate in Phase I, but install the P4007177 long rotor oil pump pickup to ensure adequate delivery (75 psi max on a B motor), as well as a stronger P3571071 distributor and oil pump driveshaft. A windage tray will help only in those areas where the oil is closer than 6 inches to the crank. Section the Chrysler 3751236 tray to cover those areas **Continued on next page**

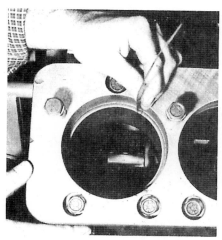

If it makes sense to hone block with a torque plate, then it's equally apparent that rings should also be gapped using a torque plate, as shown here.

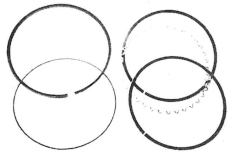

Phase II and III use Speed-Pro .005-over-size RHL-38P-9 Headland ring designed for 4.375 bore (hence reason for .055 overbore). The 3/16 diameter low-tension oil ring (SS-50U-576) locates .130 below top ring and should barely hang in bore without falling out. You may have to go through several ring sets to find eight with proper amount of tension.

MAGNUM MIRADA

only, and discard the rest of the tray.

This "mild" engine should get you into the mid-elevens in a 3000-pound car with 4.88 gears and 2500-6500 rpm power range. Not bad, but you ain't seen nothin' yet!

Phase II: NHRA-Legal Super Stock

Phase II blocks are overbored only .055-inch because of piston ring availability problems. The lifter bores should be bushed to prevent oil pressure loss if a tappet exits the bore. (This should be done on any engine running solid or roller lifters; hydraulic tappets, of course, need constant oiling.)

When the crank is index-ground, Super Stockers usually add .010-.012 to the stock stroke; that's well within the .015 stroke variance from stock permit-

ted by the rules, and helps compensate to some extent for having to run stock compression ratio pistons. It's against the Super Stock rules to lighten the crank throws.

Race-only DC P3412037 Babbit-type main bearings are preferred over Clevites. Likewise, the rod bearings are Speed-Pro Babbit (or "Micro") CB1212Ms, already drilled for rod dowel pin holes and chamfered to allow use with Hemi-style cranks having large radii journals. One "neat aspect" to using these bearings is that used ones can be lightly touched up with Scotchbrite to look like new. Paul says he's used some of the Babbits for as long four years.

Van Auken custom-built Rossi's Stage III oil pan to fit in his unique Mirada. Rear acceleration baffle sealed to rear and sides of pan keeps oil from running up into cylinders 7 and 8 and rear crank journal, is recommended for any pan (including stockers). Tube passing through pan is for steering center link; it should be only large enough to provide clearance through full lock-to-lock travel of steering linkage. Driver's side scallop was necessary to clear left idler arm.

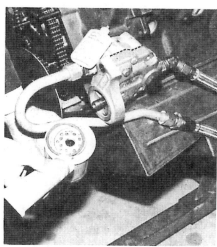

To bypass restrictive internal block oil passages, Milodon aluminum oil pump assembly (P3690039) was used. It features extra deep rotors and dual external entry for braided steel oil lines that can be purchased together with necessary fittings under P3412026. Both pump and fittings are available together from Milodon; order part No. 21185. Be sure to use Loctite on oil filter mounting stud, and "yellow death" on both oil filter gasket and pump halves (dotted line), as any failure here could be disastrous at high rpm!

Super Stock engines are required to use stock dome configuration 10.5:1 pistons, but any ring combo, skirt configuration and weight is okay. This mandates a custom lightweight piston like those made by Venolia for Rossi. The pistons are machined for double .042-inch-thick Tru-Arc retainers that retain B&B Performance .100-inch-thick 4120 steel pins. They use only two rings. Paul has been equally successful with either an .031-inch Dykes or 1/8 Headland top ring. These rings depend on maximizing gas pressure to seal effectively, so they're installed only .100-inch below the piston deck. The second ring is a low-tension oil control ring. This combo likes a smooth wall finish—use about an 825 Sunnen stone.

Lighter pistons mean lighter con rods. Paul prefers the DC P3690649, based on the old Six-Pak forging. High-strength 3/8-inch-diameter nuts and bolts are included, or they can be purchased separately under part No. P4120068.

Phase II (and III) engines need a roller cam to be competitive. Rossi likes the Crane R286/500-8, but DC's P4007279 has virtually the same specs: 286 degrees duration and .750 lift. The billet is ground on 108-degree centers and installed straight up. In Super Stock, this cam is used with the Bryant Racing Enterprises double spring, along with the DC retainer and 3-bolt roller timing chain specified for Phase I. The entire stock rocker arm shaft assembly has to go; replace it with aluminum roller rockers on a chrome-moly rocker shaft. With this setup, the stock rocker arm shaft supports are no longer adequate. Two methods to strengthen them are available: The econo approach is to replace the stock supports with aluminum support blocks made by Henry's Competition; no machining of the stock pedestals is required. Stronger (since they completely surround the shaft), but considerably more involved, is the installation of the MoPar W2 rocker arm support package (P4120102). These A-motor pieces are drilled off-center; they must be rotated 90 degrees and re-drilled on-center to work correctly with the B-motor, whose pedestal stands must be machined flush to accept them. You'll need 7/16 chrome-moly pushrods too.

No head porting or polishing is allowed! All that can be done is to bring the head down to minimum cc's and perform an NHRA-legal valve job. Valve size must remain stock (Rossi prefers Speed-Pro No. 1901 2.08-inch intakes and DC P3690709 1.74 exhausts). You'll also have to use the stock head for your year and model car. Likewise, S/S manifolding is dictated by the rules: stock carbs are required and must fit under the factory hood. Currently, the Holley Dominator (P4007937 for 383/400; P4007938 for 440) is considered to be the best 4-barrel intake with all trannies. The 440 Six-Paks use either the Edelbrock STR mentioned for Phase I, or (hood clearance permitting) the Weiand P3690982 cross-flow Tunnel Ram. Ideally, the headers' primary tube diameter must be increased from 2 to 2⅛ inches, with collector length remaining unchanged. Hooker's 2⅛ B and E body headers (part No. 5320) are 3 inches too long, but will work satisfactorily.

A more sophisticated oiling system is needed. The Milodon high-volume Hemi-style oil pump with external pickup lines and a Moroso deep pan will fit the bill. Also required is the P3690876 pump driveshaft. Specially designed for the Milodon pump and roller cam, it features an aluminum-bronze gear and hardened tip.

What does all this get you? Namely, mid-tens with a 5.12 gear. You should be competitive in SS/FA or SS/GA, using a Challenger or Barracuda.

Phase III: Super/Pro Gas & All-Out Competition

With no rules limitations, all stops are pulled out to lighten things in every feasible way. The 440 crank is turned down from its stock 2.750-inch journal size to 2.625 in order to fit in a lighter low-deck 400-inch block-bored .055-inch over. Super light custom Venolia "shorty" pistons can then be used, resulting in further weight savings. And, of course, that means an even lighter con rod—either a fully reworked standard production 440 forging (2406770) or the preferred Bill Miller aluminum rod (426002) that's much cheaper than the stock rod, costing only around $350 a set.

The Phase II roller cam and associated valvetrain components are retained, but are controlled by a gear drive set-up in place of the roller chain previously used. This higher rpm (7500 max) motor also requires Rossi to go to the ultra-heavy-duty Crane "877" (part No. 99886) triple spring and its special titanium retainer.

The trick P4120352 Stage IV heads are used. They feature enlarged intake and exhaust ports for increased top-end horsepower, with added metal surrounding each port to allow for additional grinding. The result is approximately a 2 percent flow improvement over other stock B heads (except the ultra-rare Max Wedge heads). Unfortunately, only about 4000 Stage IV heads were made and they're out of stock nationally, although ProParts still has some. If you can't get a set, "906" heads can be substituted and are preferred in this application over the '67 440 heads because the 906 open-chamber design allows a tighter piston-to-deck clearance for ultra-high compression in a milled block.

Timing chains aren't adequate for Phase III engines. Instead, use a gear drive setup like this one made by Donovan.

MAGNUM MIRADA

Stage IV porting is similar to other B heads, except everything is just a little larger—and completely polished. Mullen 2.19 intake and 1.81-inch exhaust valves with thin 5/16-inch stems are utilized for maximum flow velocity. The throat area on the valve job can get as steep as 75 degrees with the .700 + lift cams used in all-out competition, with seats .040-inch wide on the intakes and .050 wide on the exhausts. Rossi's choice of a Phase III intake hasn't yet been finalized. The initial outings of the car used the proven Weiand Six-Pak Tunnel Ram with adapter plates to mate it with the low-deck block. In line for future testing and development on Rossi's car are two 2x4 Tunnel Rams—the Weiand No. 1987 and the new Mullen 2x4 Tunnel Ram with adjustable runners and plenum. Hooker custom-made a set of headers to fit Rossi's unique Mirada; they have the same dimensions as those recommended for Phase II. Ignition is also the same, but a P3690275 adapter is needed to mate the distributor to the low-deck block. A custom-fabricated oil pan similar to that shown in the photos is recommended.

Built this way, a 3000-pound car thoroughly sorted out should do 9.50s. Right off the trailer, Rossi has already gone 9.98 @ 139 mph. It looks like he has "MoPower" than he needs to run NHRA's 9.90 Super Gas bracket. Next month, we'll discuss manifold mods, some general racer tips and see the car in action. **HR**

B-MOTOR SPECS AND CLEARANCES

Crank runout	.0015 max
Crank thrust clearance	.008
Main bearing clearance	.002-.0025
Rod bearing clearance	.003-.0035
Rod side clearance	Steel: .010-.015; aluminum: .030
Main cap torque	85 ft.-lbs.
Rod bolt torque (in oil)	Heavy-duty steel: 50-55 ft.-lbs.; aluminum: 70 ft.-lbs.
Head bolt torque	80 ft.-lbs.
Deck height	Phase I: -.045; Phase II/III: 0
Combustion chamber cc	2806762: 67cc; 2832904: 79.5cc; P4120352: 74cc
Cylinder-wall thickness	.125-.130 minimum after boring
Piston-to-head clearance	Phase I/II: Aprox. .075; Phase III: .044 (.025 quench plus .019 compressed gasket thickness)
Piston-to-valve clearance	.035-.040 intake, .060-.065 exhaust @ 0 valve lash with check springs (absolute minimum for automatics; increase slightly with sticks)
Piston-to-wall clearance	.0085-.009 (TRW or Venolia)
Valve lash	Phase II/III: .035-.040
Ring end gap	Phase I (DC ring): .016 top, .010 2nd; Phase II/III (Headland): .016 top, .007-.009 2nd compression (if used)
Valve spring pressure	DC P2806077: 115 psi @ 1.900 installed height; Bryant 8021: 145 @ 2.00; Crane 877: 185 @ 2.00
Pushrod length	Low-block: 8.700; High-block: 9.300
Spark plugs	NGK BP6S gapped at .080
Advance curve	With SoCal "Orange" racing gas: Phase I/II: 36-37 degrees @ 3000 rpm; Phase III: 35 @ 3000. Full advance in off-idle; check total timing at 3000

Building Your First Engine no. 8

INVASION OF THE KILLER B'S

Building the Big-Block Mopar

By Jeff Smith

The performance plan at Chrysler has always been very transparent: Let the big-inchers carry the ball. Oh sure, the small A-engine motors have distinguished themselves in competition both on and off the track, but when you talk Mopar muscle, your audience automatically expects you to speak the language of the 440 or the 426. While the Hemi is in a class all by itself, the 440 could be considered the grass-roots racer's powerplant. Whether you turn an occasional tire on the street or are totally immersed in the bracket scene, the 440 has to be the premier powerplant of Mopar's performance prospects with its reputation for brutal horsepower output and rock-solid reliability.

You'll notice that our supplications to the sultans of thrift are conspicuously absent here. That's for good reason. Few street machiners build 440 mileage motors. If you're going to build a 440 performance piece, you might as well make some horsepower without trying to compromise it with mileage considerations. Those mutual considerations will only disappoint you. This brings us to our 440 Mopar engine expert Marlin Gorski, who for more years than he cares to recall has been building big-inch Mopar mastodons for both street and strip combat assignments. We found ourselves in his two-car garage recently while he was assembling a bracket-bound 440 motor for an acquaintance, and we recorded some of his recommendations for building a solid street Raised Block.

BLOCK TO BASICS

Since we're going to concentrate on the RB 440 engine, we'll mention only in passing that many of the accessory pieces for the RB engine will also fit on the B-motors (with the exception of the intake manifolds). However, internally there is a whole new set of rules. If you're going boneyard hunting for a suitable horsepower foundation, look for a motorhome block with its thicker cylinder walls or at least a 1968 or later block with strengthening ribs near the freeze plugs. These blocks are somewhat superior to the older castings, which are devoid of any ribbing. Late-model 1976-'78 blocks are thinwall castings and should be avoided since they allow a maximum of only .020-inch overbore.

Of course the block should be subjected to a thorough discourse on the virtues of deburring, lifter bore honing (with a brake cylinder hone) and bolt hole tapping, along with a countersink operation just for good measure. In addition, all the freeze and oil galley plugs should be removed prior to hot-tanking. After a quick degreasing, check the block for cracks or serious maladies before having it bored (if necessary) and honed to

Cylinder head choice can have a direct effect on static compression ratios. This is the 1967 small chamber (80cc) 440 head with the larger 2.14 and 1.81-inch intake and exhaust valves. These valves are only slightly superior to the smaller stock valves. Only invest in them if you're going to replace all the valves anyway. Direct Connection Street Hemi or Sig Erson valvesprings are recommended with the cams listed in the text.

INVASION OF THE KILLER B's

TRW replacement forged pistons are the hot tip for a strong street motor. Don't let their demure appearance fool you; more than one Road Runner has run 12's with these babies. The stock 440 rod is more than adequate when outfitted with DC high-strength bolts and TRW bearings.

10-15 micro-inches for the moly rings. This should be done after the pistons have been measured to ensure correct piston-to-wall clearance.

The block should be scoured completely at least twice before you begin your final assembly steps. This cleanup procedure should commence immediately before the final assembly. The Direct Connection small parts kit will also be of tremendous assistance when it comes time to replace all those freeze plugs, dowel pins and other assorted goodies.

CYLINDER HEAD COUNSEL

Although all B and RB cylinder heads are interchangeable, there are only a few castings that are worth the

investment. The glamour, prestige and inaccessibility of Max Wedge heads aside, the best head is the 1967 440 piece with the casting number 2780915. With intake and exhaust valve diameters of 2.08 and 1.74 inches respectively and small 80cc chambers, these heads reflect the best compromise of both reasonable

The 440 could use a slight oiling system improvement with the addition of the larger ½-inch diameter Hemi oil pickup tube. This also requires drilling the stock ⅜-inch threaded hole in the block and cutting new threads.

The stock Mopar oil pump rotor kit should rejuvenate your stock oil pump along with the addition of a high-pressure black relief spring.

cylinder pressures and good port design. The next best head is the 1968-'71 larger combustion chamber version (casting number 2843906) with 89-90cc chambers. After 1971, all RB castings lost their superior port design and also succumbed to much larger emissions-dictated combustion chambers.

There is also no benefit to investing additional cash in the 440 Six-Pack heads since they are exactly the same as the '68-'71 heads. Only the addition of a good professional three-angle valve job and bronze wall guides should be necessary to bring either of the above castings up to high performance levels. If your rebuild also

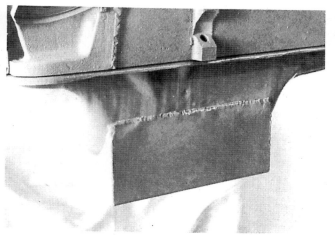

If you're contemplating some serious bracket sessions, insure your motor against the "lack of lubrication blues" by fabricating a 2-inch deeper sump. The Direct Connection racing manual #2 contains all the specifics on building such a pan.

dictates new valves, you can step up to the larger 2.14-inch Direct Connection intake and 1.81-inch exhaust valves which only require extra machining to fit either of these heads. To complete the cylinder head rebop, add either Direct Connection Street Hemi valvesprings (set at 1.870 inches installed height with 160 pounds of seat pressure) or comparable Sig Erson coils and a set of Viton rubber valve seals. And if you plan anything more than grocery-getting, immediately trash the stock rocker arms and invest

The Direct Connection small parts kit is invaluable when assembling a big-block. The kit contains all the freeze plugs, dowel pins and other assorted goodies to seal up an assembled shortblock. All DC parts can be obtained from any Chrysler or Dodge dealer. Alhambra Dodge (1401 West Main, Alhambra, CA 91801) or Norm King Dodge (121 E. Alosta Ave., Glendora, CA 91740) are two of the strongest DC dealers in the Los Angeles area.

in a set of heavy-duty rockers from the Direct Connection parts bin since the stockers are guaranteed to fail when used with the stronger valvesprings. Correct push rod length is imperative and can be accomplished with adjustable push rods available from Isky or the Smith Brothers.

PISTON RECITAL

While there's essentially nothing wrong with choosing a set of stock cast aluminum pistons, even for occasional strip blasts, Marlin likes the inherent strength of forged pistons. His long experience with the 440 race motor has led to his choice of Arias custom pistons for motors that will see almost exclusive strip duty. But even bracket motors don't need giant compression domes, and the thin 1/16-inch racing rings are definitely not the hot tip for street use. Consequently, the TRW flat top pistons and 5/64-inch stock-type rings are the hot setup for street use when you want to keep the compression down to realistic levels of 9.5:1 or below. TRW's forged replacement-type piston for the 440 makes piston choice a snap and also dictates the use of the larger single moly rings. Don't make the mistake of choosing low tension oil

control rings in an effort to "trick up" your street motor since this choice will only make the miles more unpleasant.

ROD REPERTOIRE

Building a bulletproof big-block doesn't mean expending vast amounts of cash on expensive steel rods. As a matter of fact, the stock castings (casting number 1851535), subjected

A Sig Erson TQ20 hydraulic cam is an excellent street piece when used with stock exhaust manifolds or small tube headers. A reliable timing chain is critical to cam timing and comes in the form of a Direct Connection matched set.

to a thorough inspection which includes checking their straightness, Magnafluxing and shotpeening, should prove more than adequate for anything short of all-out competition. Since you're probably going to use TRW or Sealed Power pistons, don't bother going the full-floating pin route. For competition motors this helps to equalize the deck heights by moving the bushing inside the small end, but for a street or bracket motor, it is unnecessary.

However, we shouldn't overlook a set of high-strength rod bolts from the Direct Connection clan as an integral part of a standard rod rebuild. These 3/8-inch bolts are more than strong enough to handle the torsional and compression loads as long as you can stay away from the lure of stratospheric shift points.

BEARING SOLILOQUY

Engine bearings are one of those stalwart engine pieces that, if selected properly and subjected to the proper environment, you'll never have to worry about. Marlin's experience with big-inch Mopars has taught him to select soft composition bearings that will absorb the occasional errant metal chips or

Heavy-duty rocker arms are a must on any strong big-block. Be careful when installing them on the rocker shaft. There is a left and right arm for each pair of valves per cylinder. This is the correct installation of the intake and exhaust rocker arms.

INVASION OF THE KILLER B's

The factory Six-Pack induction system is a killer on the street or strip and can also supply decent mileage if driven sanely. An Edelbrock Torker and 780 Holley or Thermo-Quad will also perform admirably.

dirt particles that might get by the oil filter. This of course goes for both the rod and main sets, and these soft bearings can be obtained merely by asking for TRW's MS2481 main and CB828 rod sets at your local parts house. Clearances for both sets are given in the specifications chart and should be checked for each rod and main. The cam bearings are also TRW pieces, but Marlin cautioned that all the cam bores should be checked with an inside micrometer to uncover any irregularities. You should pay particular attention to the number three and four cam bores (counted from the front of the motor), as these two bores seem to be the usual culprits with either out of round or undersize conditions.

PUMPING OIL

Very few lubrication changes are required on a performance Mopar to make it reliable. For average street use with an occasional bracket blast, you can leave the oiling system stock. But like all good car crafters, you can also make the factory system better. Exchanging the stock ⅜-inch oil pickup tube for a Hemi ½-inch tube will ensure the chances of supplying those precious petrochemicals to the bearings, especially at high rpm's. This change requires drilling the threaded block passage to a ½-inch hole and cutting new threads. (We should also warn you that this pickup has just been discontinued and may be hard to find.) It will work fine with the seven-quart Hemi pan, but if the urge strikes you, a 2-inch extension can be welded to the

bottom of the pan and an additional length can be added to the pickup tube. The pickup sump should be located approximately ½-inch from the bottom of the pan to prevent it from becoming uncovered, especially under deceleration. This clearance should also be checked even on a stock pan and pickup assembly.

Other recommended goodies for the lubrication checklist should include a black (70 pound) oil pressure relief spring for the oil pump in place of the stock red item that opens at 45 psi. This will work in conjunction with the standard Mopar replacement rotor kit

(for all 383-426-440 engines). Thoroughly check the housing and cover for cracks or scoring that would preclude rebuilding the pump. A windage tray is another "must run" item since it cuts down on horsepower loss at higher rpm's. This tray does require a pair of pan gaskets since it mounts between the pan and the block, so be sure to get an extra pan gasket along with the tray.

CAMSHAFT COMMENTARY

Since any street engine can be

though stout compression ratios of 11 or 12:1 sound impressive, they can create a nightmarish detonation destruction derby on today's weak excuse for gasoline. Keeping it below 9.5:1 will save you this headache, and the engine should run on almost any kind of good pump gas. Listed below are a few different compression combos that are possible with only flat top pistons and the right deck heights.

CLEARANCES (Inches)

Main	.002-.0025 (optimum)
Rod	.002-.0025 (optimum)
Piston-to-wall	.002 (TRW) minimum
Rings	Top—.018 Second—.010-.012
Valve-to-piston	Intake—.100 Exhaust—.120
Side (rods)	.009-.017
Crankshaft end play	.004-.009

TORQUE SPECS (Ft./Lbs.)

Head	70
Main	85
Rod	45-50
Crankshaft snout	135
Flywheel	55
Intake	40
Exhaust (stock)	30

THE COMPRESSION COMPROMISE

There are literally dozens of different static compression combinations that are possible with the big-block Mopar. By combining different pistons with different deck heights, gasket thicknesses and combustion chamber volumes, you can tailor the specific compression ratio you wish to run. We should warn you that, al-

	.020-inch HEAD GASKET	.035-inch HEAD GASKET
Cylinder head (casting 2780915), 80cc chamber		
	11.0:1	10.6:1
Piston—TRW L2310F w/-.022-inch deck and standard bore		
Cylinder head (casting 2843906), 86cc chamber		
	9.38:1	9.1:1
Piston—TRW L2266F w/-.077-inch deck and standard bore		

Big-Block Mopar

Be sure to install a paper intake gasket on each side of the tin intake valley cover. Neglecting this simple step will prevent the intake manifold from sealing and will produce an avalanche of vacuum leaks.

enhanced or destroyed by camshaft selection, the type of use (or abuse) you have in store for the RB engine should weigh heavily on any decision concerning camshaft timing. Marlin has found that the 440 Six-Pack cam has a little too much timing to make it compatible with stock exhaust manifolds or small tube headers, especially in a street-driven B-body with a TorqueFlite. Depending on how you plan to use the 440, the biggest cam for bracket bombing would be a Racer Brown SSH-25 (or the comparable DC P4120235 cam and lifter package) hydraulic cam in mandatory conjunction with an 11-inch converter and 2-inch headers. A good street application would be a Sig Erson TQ20 hydraulic with 268 degrees advertised duration that would work well with stock exhaust manifolds or headers with a primary pipe diameter up to 1⅞ inches. We're giving header and torque converter recommendations along with cam possibilities here since the big-block Mopars seem to thrive on a well-matched cam, exhaust system and converter combination, and these are a few of the permutations that Marlin has had luck with over the years.

BREATHING EXERCISES

Even though this engine buildup is rapidly approaching a stout piece, restraint is still the password here, and this goes for intake manifold selection as well. The Edelbrock Torker is well suited for both low-speed throttle response and mild rpm duties and excels when mated to either a Carter Thermo-Quad or Holley 780 cfm vacuum-actuated carburetor. If the attraction of three 2-barrels has you entranced, the aluminum intake and Holley package will also make the 440 stand up and take notice. Edelbrock produced the original 1969 aluminum casting for the Mopar Six-Pack setup, which can be obtained new from almost anywhere. The carburetors can be obtained through a number of mail-order outlets or directly from Holley, but if you are looking for an original set, try to find the later 1970-'71 carbs as they offer slightly better reliability. The DC folks also offer both jetting and tuning kits for the 6-barrel system that could be useful in fine-tuning your combination. The DC racing manual #2 also gives jetting recommendations that might be helpful.

While this segment of "Building Your First Engine" has been more of a parts recommendation outline than a "how-to" treatise, the basics of assembling an engine have been covered in previous installments.

For the weight conscious, this aluminum water pump housing is still available from the DC people under PN P2536086.

The completed 440 stormer, ready for some heavy street action.

Besides, regardless of what is stamped on the valve cover, there is only one way to build an engine correctly and that is to take your time, clean everything to the max, and preassemble the motor to check clearances until you are satisfied that everything is right on. The are no magic formulas for making an engine work for you. The only magic is in correct parts selection and pragmatic engine assembly techniques. But now that you've got the straight skinny on what it takes to build a muscle-bulging big-block Mopar, what are you doing just sitting there? Get busy! ◧

PARTS LIST

All parts are Direct Connection items unless otherwise identified.

L2266F	Piston, forged aluminum replacement, TRW
MS2481	Bearings, main set, TRW
CB828	Bearings, rod set, TRW
R-9224	Rings, single moly type, Sealed Power
P4120068	Bolt and nut set, connecting rod, ⅜-inch high-strength
SSH-25	Camshaft, bracket-competition, Racer Brown
P4120235	Camshaft and lifters, same as above, Direct Connection
410121	Camshaft, street type, Sig Erson TQ20
915031	Valvesprings, all applications, Sig Erson
P3690933	Valvesprings, Street Hemi type
P3690708	Valve, intake, 2.08-inch
P3690709	Valve, exhaust, 1.74-inch
P4007942	Valve, intake, 2.14-inch
P4120579	Valve, exhaust, 1.81-inch
P3412067	Retainers, aluminum
P4120618	Valve stem locks, one-groove
P4120620	Valve stem locks, three-groove
P3690712	Rocker arm, heavy-duty, right side
P3690713	Rocker arm, heavy-duty, left side
P3690277	Camshaft sprocket, one-bolt
P3690278	Camshaft sprocket, three-bolt
P3690279	Chain, roller
P3690280	Crankshaft sprocket
P4120492	Seals, valve stem, Viton type
P4007943	Small parts kit, B-RB engines
3751236	Windage tray
P4007819	Rotor kit, oil pump, all B-RB engines
2406677	Spring, oil pump, high pressure (black)
3514186	Gasket set, intake manifold
P4007039	Fuel pump, mechanical high performance
P4007671	Carburetor tuning kit, 440 Six-Pack
P3571071	Oil pump driveshaft, hardened tip
P4120792	Racing Manual, 1980 Revised Edition

MO' MONEY TO SAVE, MO' POWER TO BURN

By David Freiburger

Budget Engine SPECIAL

According to Chrysler, more than three million 383s and ¾ million 440s were produced between 1959 and 1978, when big-block production ended. Chevy seems to pump out that many small-blocks in a year. Because Mopars aren't as common, there aren't many sources for complete low-buck engines, so you'll probably have to build your big Mopar yourself. To help out, we'll tell you about the best Mopar parts sources with quality rebuild kits ready for action. And since the boneyards are filled with prime Mopar engine goodies, you'll also learn to identify the best core engines and parts. While we're at it, we'll show that a big Mopar is a better deal than a big-block Chevy! But do us a favor and keep it quiet; we don't want people to catch on and start buying up our limited supply of engines.

Big-Block MOPAR

MOPOWER IS BUDGET POWER

Some argue that a big-block Mopar is not only cheaper than a Rat-motor Chevy, but that it also offers better reliability and torque-per-dollar. Let's see why.

The mechanical advantages begin with the Mopar's large lifter diameter, which allows camshafts with a faster rate of lift that would require a roller cam in a Chevy. (Hughes Engines and McCandless Performance are two companies that offer cams to take advantage of that fact.) Second, the Mopar's rod ratio (1.89 for a 383; 1.80 for a 440) is much better for drag-racing power than the 454 Chevy's poor 1.53 ratio. Finally, the 440 has bottom-end strength beyond compare; forged cranks for 440s are easier to find than for Chevys; and even cast-crank 440s are good for 6000 rpm.

Now for the financial blow. We checked gasket kits, piston kits, and crank kits from PAW, RHS, and Speed-O-Motive. Mopar parts are 10 to 30 percent cheaper than the big Chevy and Ford kits! Mopar accessory parts are more expensive, but add up the cost for the total package. The long-block can be built for less than a Chevy; the car is cheaper if you don't "need" an R/T or a 'Cuda; it'll never break; and you look better while going faster. MoPower is affordable power.

Big-block Mopar engine kits are cheaper than big-block Chevy kits! Speed-O-Motive gets $595 for this Super Master Kit for a 440, but the price for a 454 is $665.

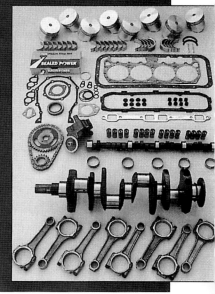

THE SCREAMINGEST OF DEALS

We located some shops that still have New Old Stock (N.O.S.) short-blocks, cranks, and rods—and they're cheap, too! Chrysler Performance Parts Association (CPPA) has new cast-crank 400 short-blocks for $400 and 440s for $900. McCandless sells cast-crank 400s for $500, 440s for $750, and even has rare forged-crank 400s for $650. Some N.O.S. short-blocks are thin-wall castings without a camshaft, but Mancini Racing has thick-wall 1972 440 bare-blocks for $400.

New bottom-end parts are also available. Cast 440 cranks are $50 from McCandless, and CPPA charges $100 with free shipping. A properly machined and balanced cast crank is good to 6000 rpm but may be cost prohibitive if you're replacing a forged crank because you'll need the externally balanced dampener and torque converter or flywheel to go with it. Cast cranks can be internally balanced, but mallory metal is also expensive.

If you need a forged crank, McCandless sells 383 and 440 forgings for $450, and new steel 440 rods are $30 a pop. You can also get remanufactured forged cranks and rods in balanced crank kits from Speed-O-Motive for $483 with cast pistons or $581 with forged pistons.

KNOW YOUR WEDGIE

Since the best deals are on assembly kits, you'll save money by hunting the boneyards for a core fat-motor to rebuild. But a great deal on a "'70 383 Magnum" is a lot less attractive when you find out it's actually a 361, so it pays to know what you're looking for. If you can't find the engine size stamped on the block, identify it by the type, year, and bore listed on the accompanying chart. The low-deck B engines measure 9.98 inches from the crankshaft centerline to the top of the block, while raised-deck RB engines are 10.725 inches from crank to top.

B-engine crankshafts are recognizable by a 3.38-inch stroke, while RB cranks have a 3.75-inch stroke. B-engine cranks have a 2.625-inch main-journal diameter; RB's are 2.750-inch; and they both have 2.38-inch rod journals.

DISPLACEMENT	TYPE	YEARS	BORE	STROKE
350	B	'58	4.06	3.38
361	B	'58-'66	4.12	3.38
383	RB	'59-'60	4.03	3.75
383	B	'59-'71	4.25	3.38
400	B	'71-'78	4.34	3.38
413	RB	'63-'73	4.18	3.75
426W	RB	'63-'66	4.25	3.75
440	RB	'66-'78	4.32	3.75

B engines have a small pad under the distributor on the right-hand front of the block (arrow A), while RB engines have it on the left-hand side (arrow B). Engine displacement will be stamped there unless you've found a '68 to '71 Hemi. Cast-crank engines usually have "E" stamped onto the pad as well.

GIT YER HEAD STRAIGHT

For a low-buck street engine, ported stock iron heads are the way to go. If you want to rebuild the heads yourself, stick with 915 closed-chamber or 906 and 452 open-chamber castings with 2.08/1.74 valves. Port them at home with the Mopar Performance templates (part No. P4120437) and you'll have the best compromise between power and budget.

Outright head prices range from $450 for a basic rebuild to $1665 for maxed-out porting and are available from CPPA, Hale, Hughes, Landy, McCandless, PAW, and RHS. Some will rebuild your heads; Hughes Engines will Pro Port, mill, Magnaflux, and add 2.14/1.81 valves and bronze guides for $695.

Most Mopar-specific shops offer ported head assemblies like this $1395 set from CPPA (springs and retainers are included). Some prefer the 452 castings because they have hardened exhaust seats for unleaded gas.

MOPAR-TICULARS

For cheap high compression, use affordable flattop pistons and '67-and-earlier heads with 70cc to 78cc closed chambers (left) rather than the later 80cc to 90cc open chambers (right). The 1967 2780915 castings are the best affordable closed-chamber heads, but you may have to upgrade the exhaust-valve seats (MP part No. P4452189 for 1.74-inch or 1.81-inch valves) if you think you can run unleaded gas, which you probably can't.

You can save money by junkyard shopping and still get the best core parts if you know what to look for. Here are some tips for spotting the best blocks, rods, and cranks. (See the photo and "Git Yer Head Straight" for cylinder-head tips.)

Starting with blocks, avoid '76 to '78 thin-wall castings because they shouldn't be overbored more than .020-inch. Also, the shops we talked to said that the musclecar blocks with extra strengthening ribs on the side are fine but aren't worth paying extra for.

As for cranks, the best 440 units are the '72-and-earlier forgings; mid-'73-and-later 440s had cast cranks. Low-block cranks were forged prior to '71 when cast cranks were introduced in 383 two-barrels. All '72 to '78 B motors have cast cranks except some '72 to '74 400 four-barrel engines.

Reconditioned standard rods are the cheapest and best choice for street use (B and RB rods are not interchangeable). Six Pak rods add more detrimental weight than valuable strength, but if you must have them, they'll be cheaper in the 440-3 motorhome and police-car engines than from the '70 and '71 six-barrel motors. Six Pak rods also require a different vibration dampener and flexplate or flywheel. The 426 Hemi forgings are actually the strongest stock steel rods and can be used in a 440 if you rebalance the crank.

NO ASSEMBLY REQUIRED

Mopar shops don't usually offer off-the-shelf low-buck engines because demand is low and each customer wants something different. However, CPPA, Hale, Hughes, Landy, and McCandless will all build budget engines to your specs. Pete Figueroa at CPPA charges $1000 plus parts and machine work for a complete carb-to-pan engine, and they're pretty enough for a cover shot.

For out-of-the-box long-block Chryslers, Racing Head Service (RHS) offers base High Energy 440s for $2895. That's an even better deal when you find out that a similar 454 Chevy is $2995!

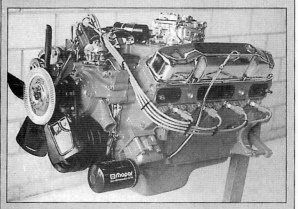

CAST IN IRON

Buying the right used MoParts is easy if you can decode the casting numbers. The guru of Mopar casting numbers is Galen Govier, who publishes a parts-interchange book that lists numbers for rods, blocks, heads, manifolds, and harmonic balancers all the way to distributors and carbs. Three production-option-code books for '66 to '68, '69 to '71, and '72 to '74 are also available. To order any of Galen's books, send $15 to G.T.S., Dept. HR05, Route 1, Box 322K, Prairie du Chien, WI 53821, 608/326-6346. **HR**

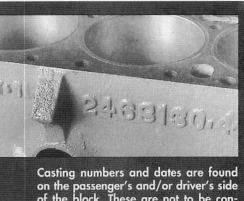

Casting numbers and dates are found on the passenger's and/or driver's side of the block. These are not to be confused with the engine serial number and Vehicle Identification Number, which are elsewhere on the block.

Head-casting numbers can be found on the underside *(shown)* or under the valve cover on an intake runner near the rocker-shaft pedestal.

SOME ASSEMBLY REQUIRED

Not many companies offer complete, assembled, mail-order Mopars, but everything from gasket kits to short-block kits are easy to come by. The combination of parts varies greatly, but all the companies in the source box have great deals. At $1150, Speed-O-Motive's Super Short Block kit *(shown)* is one of the best bargains we found on a complete short-block. It includes everything shown plus a Fel-Pro gasket set. The crank and pistons are forged, and you have your choice of cam grinds. Best of all, it's only $55 more than Speed-O-Motive's small-block Chevy kit.

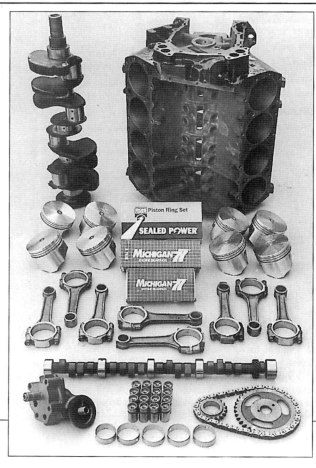

SOURCES

Chrysler Performance Parts Association (CPPA)
Dept. HR05, P.O. Box 1210
Azusa, CA 91702, 818/303-6220

Dick Landy Industries
Dept. HR05, 19743 Bahama St.
Northridge, CA 91324, 818/341-4143

Hale Performance Warehouse
Dept. HR05, P.O. Box 1518
Van Buren, AR 72956, 501/474-5252

Herb McCandless Performance Parts
Dept. HR05, P.O. Box 741
Graham, NC 27253, 919/578-3682

Hughes Engines, Inc.
Dept. HR05, 23334 Wiegand Ln.
Washington, IL 61571, 309/745-9558

Mancini Racing
Dept. HR05, P.O. Box 239
Roseville, MI 48066, 313/294-6670

Mopar Performance
Dept. HR05, P.O. Box 215020
Auburn Hills, MI 48321, 313/853-7290

Performance Automotive Wholesale, Inc. (PAW)
Dept. HR05, 8966 Mason St.
Chatsworth, CA 91311, 818/998-6000

Racing Head Service (RHS)
Dept. HR05, 3416 Democrat Rd.
Memphis, TN 38118, 800/333-6182

Speed-O-Motive
Dept. HR05, P.O. Box 4308
Santa Fe Springs, CA 90670, 310/945-3444

One Bad B

TICO Racing strokes the 400 for 502ci of street Mopower

by E.F. Nowak
photography by Jen Wainwright and E.F. Nowak

Going for maximum power-per-dollar? The hands-down advantage goes to the stroker engine. Though streetable stroked engines tend to make only slightly more high RPM horsepower than their stock displacement counterparts, the real value lies at the low end of the power range where crucial torque is made. It's not uncommon for long arm engines to match the amount of added cubic inches with an equal share of torque.

Engine torque is the single most important factor related to street and bracket racing performance. Torque is not about how much further an engine can rev to. It's about raw, gut wrenching, stab the gas and hang on for life, power. Power like that allows an engine to use lower numerical gears, like 3.55s or 3.23s, and still obtain improved vehicle performance. In addition to that, the lower gears, combined with a stroker's inherent reduced RPM band, can even extend engine life or component longevity.

TICO Racing's B-engine big-block stroker is an engine that, at 502 cubes, handily fits into several applications. As easily as it can see duty as a King of the Street engine, it also fits the need of anybody looking for a genuinely big engine. Because of its broad torque range, and relatively high HP figures, this engine can do the job in mud truck racing, pulling, street rod, bracket racing, and even some circle classes. The word universal is an understatement!

Either B (low blocks) or RB (raised blocks) can be used for big inch engines; TICO Racing generally favors the low block. Its lower deck allows it to nestle into A-Body engine bays easier. It's lighter than an RB, and because it's slightly stiffer, it offers superior cylinder sealing. TICO recommends early- to mid-'70's low blocks with improved coolant flow passages shaped like sideways 8s. (continued)

Before sending any block into the Sunnen CV-616, TICO inspects the block for cracks and wear patterns. If everything checks out, a deck plate is torqued to the block and the block gets power-honed. Power-honing with a deck plate simulates engine block torque stress. Blocks machined in this fashion will generate more power. Before assembly, TICO recommends that all the bolt threads be chased with taps to insure correct torque values.

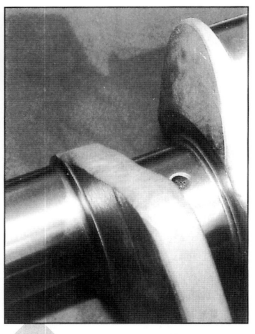

TICO Racing welds customer 440 core cranks to 4.125-inches. (See sidebar, "Mr. Tipz Sez..." on page 23 for details). Quality workmanship, stronger radiused journal ends, and less rotating weight are the primary benefits to the welded cranks.

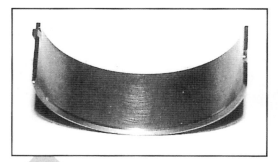

Specialty cranks with radiused journals will need a corresponding bearing to maintain bearing edge clearance. Clevite makes a performance bearing that provides clearance and durability. The part numbers are: RB rods CB-527HD, B rods MS-1277HG and MS-876P. Always check for radius-to-bearing clearance before final assembly.

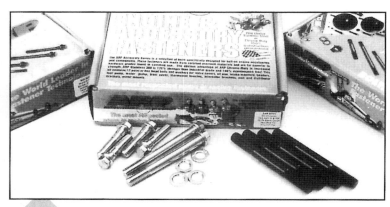

Increasing the performance level of any engine places more stress on the fasteners. ARP studs used on the main caps and cylinder heads provide repeatable true torque readings. This insures proper clamping pressures—a critical aid to true high performance engines. ARP offers complete engine fastener kits for all Mopars.

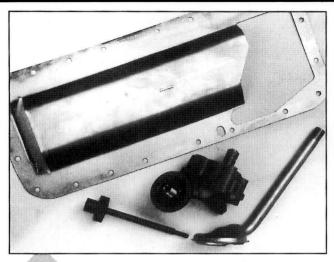

After the oil passages are drilled out, a Melling High Volume pump (M-63-HV) and heavy duty intermediate shaft (IS-63) is bolted up. A Mopar Performance windage tray (P4120998) and Hemi pick up (P4529566) complete the upgrade. TICO recommends checking rod to tray clearance on all stroker engines. In a worst case scenario, the tray may need to be split on the side and a section welded in to gain room for the rods to spin around without metal-to-metal contact.

TICO uses 440 (preferably 4-bbl LY) rods for both B and RB blocks. Standard procedure is degreasing, Magnafluxing, shotpeening, and big-end re-sizing with ARP rod bolts. The use of longer 440 rods in a low block is more desirable because they prevent the rod ratio from decreasing numerically. Poor rod angles cause more internal preloading against the cylinder walls and tend to make peakier torque curves. For the hardcore enthusiast, TICO converts the pin to a floater by reducing the small-end size down to .990 inch (from standard 1.09 inch). This upgrade lightens the reciprocating weight by over 100 grams per rod and must be accompanied with a custom-made piston.

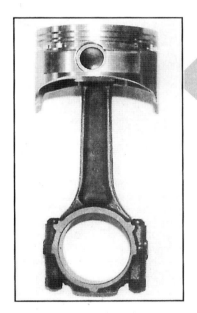

Ross Racing Pistons are custom-made per application, super strong, and exceptionally lightweight. It's not uncommon to get the balance factors down to near small-block weights with the Ross/LY rod/.990 pin combination. Ross Racing Pistons can machine any compression ratio specified by the builder. This particular build was targeted at the street and the comp ratio was held to 9.5:1.

Building a successful deep cylinder bore engine lies in the cylinder head flow. Like the block, the heads are boiled, inspected for cracks, and the guides cut for high lifts and PC seals. TICO Racing uses induction-hardened 452 heads, or 906s with hardened seats installed.

Though the valve combination seems to run contrary to popular beliefs, TICO has had considerable success by deploying REV Hi-Temp 21-4N forged stainless steel 2.14/1.74-inch valves, part numbers 3556E and SEV2665SG, respectively. Ported, the heads flow an almost unheard of 75 percent (@ .500-inch lift) balance ratio. That's on par with many out-of-the-box aftermarket aluminum heads. (See "Cylinder Head Flow Comparison" chart on page 22 for flow figures).

Cylinder Head Flow Comparison

Valve Lift (in.)	Stock Intake (cfm)	Stock Exhaust (cfm)	Stock Balance %	TICO Intake (cfm)	TICO Exhaust (cfm)	TICO Balance %
.100	58	38	66	78	61	78
.200	130	89	65	153	118	77
.300	175	119	68	211	161	76
.400	202	129	64	243	185	76
.500	210	129	62	261	195	75
.600	210	129	61	268	196	73

Stock 906 Casting: 2.08/1.74 valves.
TICO 906 Casting: 2.14/1.74 REV valves.

502 DYNO FIGURES

Engine RPM	Horse-power	Torque (lb-ft)
2500	229	480
3000	284	497
3500	346	520
4000	411	539
4500	471	**550**
5000	511	536
5500	**528**	504
6000	519	454

Without matching the intake's flow potential to the ported cylinder heads' flow, a stroker engine will fall flat at the upper levels of the RPM range. Weiand's Team G offers runners straight as an arrow and a plenum size that's right on the money for serious high-performance engines. Though the dimples at the end of the runner do little to restrict flow into the head, TICO will touch the area with a grinder and then do some minor polishing. Use PN 7533 for low blocks and PN 7534 for raised blocks.

TICO Racing has tried a number of different carbs. Holley's manual choke 850 cfm carb (PN 0-80781, left) is the street ticket. The 502ci headed for track use will find the high-velocity, super-responsive 950 cfm (PN 0-80497, right) ideal. The 950 carb has a horsepower advantage of 10-15hp at the upper end of the power curve, but tends to be hard to live with on the street because there's no choke for cold starting. In either case, both of these carbs feature jets in all four corners (instead of a metering plate at the rear) for fine tuning and maximum power production.

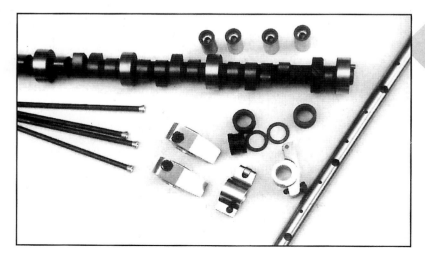

Wolverine Blue Racer presents the builder with some pretty unique street/bracket cam profiles and outstanding valve gear. Wolverine's (WG-1067K, .495-inch intake/.518-inch exhaust lift, and 301/305 degree duration) solid lifter grind provides enough duration to fill the long arm engine's cylinders without sacrificing vacuum needed for bottom-end grunt. The mid-range lift figures of the cam, combined with the 1.74-inch REV valves, eliminates potential valve/piston clearance problems without sacrificing horsepower or torque. Wolverine also offers the ultimate shaft rockers for the B/RB engines. The precision aluminum 1.5:1 ratio, roller-tipped rockers (WG-6018-16) and strong .080-inch moly push rods insure valvetrain stability at high RPM. Rocker shaft is a stout Melling piece.

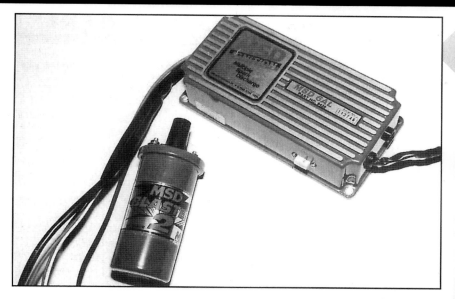

To insure complete combustion, TICO uses an MSD-6AL and Blaster 2 coil. The 6AL features a wide range of RPM-limiting chips, not to mention multiple sparks at the firing cycle, making tuning a breeze.

MR. TIPZ SEZ:
Shoehorning A 440 Crank Into A Low Block Is Easy

If yer sold on doin' a stroker 400, here's the scoop on how to drop a 440 crank into the low block. Now, we know the 400 block won't just bolt up a 440 crank 'cause the mains are different sizes, the Bs being 2.685 inches and the RBs being 2.75 inches. So, first off folks, let's go over "the wrong way." That's to bore the low block's main webs out to larger RB sizes. Bad move, because it will make the main caps weak, and it's more expensive than turning down the crank's mains.

Actually, the whole conversion is rather straight forward and relatively easy to do. You start out by turning the mains to standard low-block diameters (A), then have TICO grind a radius at the journal ends instead of roll them like the weaker stocker cranks. Next, you'll

need to turn down the 440 crank's counterweights by .250 inch (B). Then, notch the block's lower cylinders (C) and trim back the oil pick-up boss (D) so the rods will clear those respective areas. The last two features will allow you to weld up your stock 440 crank to 4.125 inches (TICO Racing can do this for you) like the one done in E.F.'s story.

So, why do this if the aftermarket offers non-welded, billet cranks? Them aftermarket jobs are heavy, so the engine will tend to rev a mite slower. Besides that, the heavier cranks will require even heavier Mallory metal to be added just to get the balance job right, making the whole enchilada much more heavier. On top of that, the heavier a crank gets the more it will be inclined to pound out the bearings.

Using a lighter factory 440 crank will keep the weight of the rotating mass at a minimum. And, chances are that if you use 440 LY rods, you won't even need any Mallory to straighten out the balance job.

Now, when we look at the bottom line, you're going to find that the welded-up 440 crank is going to be a few hundred dollars cheaper than the billet crank. If you're not making over 550/575hp, the welded-up stock steel 440 cranks work just fine.

SOURCES:

AE Clevite
Dept. MPRM
1350 Eisenhower Place
Ann Arbor, MI 48108-3282
(800) 338-8786, ext. 528

ARP
Automotive Racing Products
Dept. MPRM
531 Spectrum Circle
Oxnard, CA 93030
(800) 826-3045

Holley Performance Products
Dept. MPRM
1801 Russellville Rd.
Bowling Green, KY 42102-7360
(270) 782-2900

MSD Ignition
Autotronic Controls Corp.
Dept. MPRM
1490 Henry Brennan Dr.
El Paso, TX 79936
(915) 857-5200

REV
Racing Engineered Valves
Dept. MPRM
4704 NE 11th Ave.
Fort Lauderdale, FL 33334
(800) 398-6348

TICO Racing Engines
Dept. MPRM
10600 Chicago Rd (US 12)
Somerset Center, MI 49282
(517) 688-8426

Weiand Automotive Industries
Dept. MPRM
2316 San Fernando Rd.
P.O. Box 65301
Los Angeles, CA 90065
(323) 225-4138

Wolverine Blue Racer
Dept. MPRM
4790 Hudson Rd.
Osseo, MI 49266
(800) 248-0134

MOPAR HEAD COMPARO

BY MARLAN DAVIS

Photos and graphics by Marlan Davis

For over two decades, Mopar enthusiasts have relied on '68-'70 440 cast-iron, high-performance cylinder heads for their mainstream big-block B-engine performance cylinder head requirements. These classic castings are commonly referred to as "906" heads, an identifying shorthand for their complete "2843906" casting number. While 906 heads were good all-around heads in their day, that day has passed—it's been 30 years since these heads were factory-installed on a production car. Castings in good condition without multiple valvejob-induced sunken seats are scarce. And none of the originals have unleaded-gas–compatible hardened exhaust valve seats.

Although today both Mopar Performance (MP) and aftermarket sources offer modern B-engine aluminum heads of varying sophistication—including modified B1 castings suitable for use in NHRA Pro Stock competition—most of these heads are not direct bolt-ons. They may feature raised or relocated ports, require dedicated intake manifolds, and/or utilize a unique valvetrain. In other words,

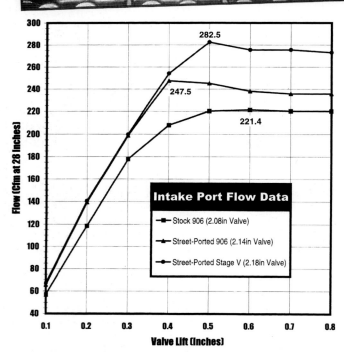

Intake Port Flow Data

- Stock 906 (2.08in Valve)
- Street-Ported 906 (2.14in Valve)
- Street-Ported Stage V (2.18in Valve)

282.5
247.5
221.4

Flow (Cfm at 28 Inches) vs Valve Lift (Inches)

Street-porting the 906 and Stage V heads yields significant flow benefits in comparison to the unported baseline 906 stocker. The Stage V's greatest intake-side improvements occur at over-0.400-inch lifts.

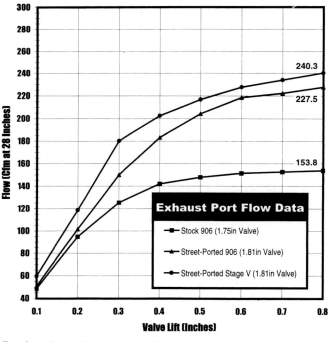

Exhaust Port Flow Data

- Stock 906 (1.75in Valve)
- Street-Ported 906 (1.81in Valve)
- Street-Ported Stage V (1.81in Valve)

240.3
227.5
153.8

Flow (Cfm at 28 Inches) vs Valve Lift (Inches)

Porting the 906's exhaust side offers a huge improvement over the unported baseline. The ported Stage V offers further enhancements, but the added incremental improvement—while still substantial—is not nearly as much as the intake-side gains.

they'll set you back a pretty penny.

Recognizing a need for affordable, bolt-on replacement cast-iron performance B-engine heads, MP offers Stage V cast-iron cylinder heads (P4529993). At first glance, they look identical to the old 906 castings, but the interiors of the new heads' runners are subtly recontoured with improved short-side radii. According to the Mopar specialists at Dick Landy Industries (DLI), both the old 906 castings and the new Stage V replacement heads respond well to typical street porting, but the Stage V's airflow improvement is incrementally greater.

Go With the Flow

To illustrate the flow differences, Landy first tested a pristine (no sunken seats) 906 head with stock-type 2.08-inch intake/1.75-inch exhaust valves. These heads' intake essentially stalls at 0.500-inch lift (see graph), with the flow at 28 inches peaking out around 220 cfm. On the exhaust side, flow above 0.500-inch lift barely exceeds 150 cfm.

Treating the 906 castings to Landy's street-porting regimen—larger 2.14-inch intake/1.81-inch exhaust valves, bowl work, pocket porting, a five-angle valvejob, overall port cleanup, and gasket match—shows healthy 11.8-percent peak and 11.3-percent average intake flow gains, but it's the exhaust side that really shines: Peak flow increases by a whopping 47.9 percent (from 153.8 cfm to 227.5 cfm at 0.800-inch lift), with overall average flow numbers showing a 33.6 percent improvement (from 127.1 to 169.7 cfm).

Landy next applied the same porting techniques to Stage V castings, which were also enhanced with bigger 2.18-inch intake valves used in most Landy-ported B-engine bracket-racer heads (they retain the ported 906's 1.81-inch exhaust valve). Although at first glance the 2.18 valves may seem to muddy the comparison waters, Landy claims that, if anything, the huge intake valves' shrouding effects actually hurt low-end flow. The flow tests bear this out, with the majority of intake-side gains occurring beyond 0.400-inch lift. Intake flow now peaks at 0.500-inch lift, where Landy's SuperFlow bench recorded 282.5 cfm, a gain of 14.1 percent peak-to-peak over the ported 906 head, and 27.6 percent over the unported 906 stocker. The Stage V's average 221-cfm intake flow numbers represent 22.5- and 10.1-percent improvements over the stock baseline and ported 906 head configurations respectively.

With the same size exhaust valve,

Stage V flow is up throughout the 0.100-0.800-inch-lift test-range, but the percentage improvement compared to the ported 906 head—5.6 percent more flow at peak lift, 8.9 percent on average—is not nearly as high as the intake side's gain. Compared to the stock, unported 906 casting, the Stage V exhaust shows 56.2 percent more peak flow; at 184.9 cfm, its average flow is up by 45.5 percent.

Dyno-Might

Raw flow bench data is one thing; how (and if) that translates into improved engine output is another. Landy tested the ported 906 and Stage V heads on the

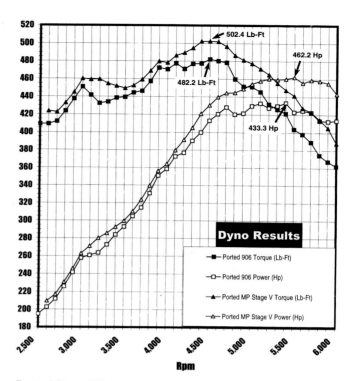

Ported Stage V heads help make over 450 lb-ft between 3,000 and 5,300 rpm, a 500-rpm–wider torque band than was developed by ported 906 heads. Peak-to-peak, the Stage V heads are also worth nearly 30 hp more.

same mild 0.030-inch-over 440 street engine. The 9.0:1 motor was equipped with a relatively mild DLI BH-228 hydraulic flat-tappet cam that specs out at 0.480-inch intake/0.502-inch exhaust valve lift (with 1.50:1 ratio rockers), 284 degrees intake/294 degrees exhaust advertised duration (228/238 degrees at 0.050-inch

Old 906 (*left*) and new Stage V (*right*) heads are externally identical, but current Stage V castings have a functional heat riser and are legal for service replacement and NHRA Stock and Super Stock use.

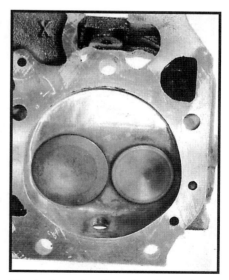

Nominal combustion chamber volume is set to 79.5 cc. Both of the ported heads used for testing had Landy's $90 "bracket cc"—all four chambers are checked, then the heads are milled to achieve the desired overall average chamber volume.

The best bang for the buck, DLI's $1,495 "street-port" includes short-side radius work, a gasket-match, a multiangle valvejob, and general runner cleanup.

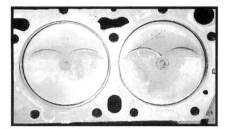

DLI's mild short-block is stuffed with a stock forged crank and stock rods fitted with ARP bolts. Federal Mogul's L2266F-30 flat-top forged pistons were lightened and notched for valve clearance (mods not really needed for this test); Landy also pin-fit the pistons to the rods.

Equipped with a nonadjustable valvetrain, DLI's engine uses stock pushrods, heavy-duty 1.5:1 ratio non-adjustable rocker arms (MP PN P4529743, package of 16), P4529101 rocker shafts, new P5249848 chrome-silicon springs (safe through 0.540-inch lift), P4452033 chrome-moly retainers, and P4120618 8-degree locks.

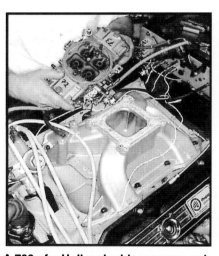

A 700-cfm Holley double-pumper carb (No. 69 primary/No. 78 secondary jets, 6.5 power valve) was used on a standard Mopar M1 single-plane intake manifold that, says Mike Landy, "wasn't even port-matched."

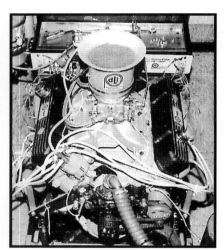

Run on 92-octane street gas, the 9.0:1 engine was equipped with a Mopar electronic distributor (36 degrees total advance) triggered by an MSD-6 box and Hooker headers with 2-inch primaries and 12-inch-long, 3.25-inch-od collectors.

tappet lift), 112 degrees lobe displacement angle, and 107 degrees installed intake centerline.

As shown by the "Dyno Results" graph, the ported 906 heads developed 482.2 lb-ft of torque at 4,500 rpm; their peak power output of 433.3 hp occurred at 5,400 rpm. Swapping the heads for the ported Stage V castings gained 20.2 lb-ft and 28.9 hp at the peaks, for a total peak output of 502.4 lb-ft at 4,500 rpm and 462.2 hp at 5,500 rpm. At the peaks, this represents a 4.2-percent torque and a 6.7-percent power increase; on average, the Stage V heads are worth about a 5-percent power gain and a 4.7-percent torque increase, with the majority of the improvements occurring over 4,200 rpm. According to Landy, the lower improvements achieved on the actual engine in comparison to the flow bench numbers are typical for a real-world street engine of this type.

The Price of Performance

Naturally, the extra performance doesn't come free. While exact costs may vary, if you had to go buy a used set of 906 castings from a salvage yard, figure on shelling out around $100-$400 for a pair of cores in uncertain condition. Add cleaning, Magnaflux inspection, street-porting, modern-day valve seal assemblies, and

new stainless valves (all done by Landy), and the bill comes to about $1,795.

Landy sells new bare Stage V castings for $1,095. Add in the same degree of preparation, and Landy's out-the-door price comes to $2,195 (including the castings). In other words, you gain nearly 30 hp for another $400. That's about $13/hp on the test mule-motor. Landy reports that gains would be greater on big-cam, higher-rpm engines, and you'd have the peace of mind of having new, stronger parts instead of worn-out, thrashed, cobbled-together antiques. **CC**

SOURCES

DICK LANDY INDUSTRIES
Dept. CC
19743 Bahama St.
Northridge, CA 91324-3366
818/341-4143

MOPAR PERFORMANCE HEADQUARTERS
Dept. CC
P.O. Box 360445
Strongsville, OH 44136-9919
Catalogs, tech manuals: 800/348-4696
Tech line: 248/853-7290
www.mopar.com

"A" MOTOR RECIPES

DIRECT CONNECTION'S PLANNED INGREDIENTS GET YOUR SMALL-BLOCK MOPAR COOKIN'!!

By Marlan Davis

MOPAR "A" MOTOR
1654568
1263274
21481537
3109468

There are two ways to bake a cake. The "half baked" method, also known as the trial and error approach, usually ends up being quite a trial with lots of error. On the other hand, you can seek expert advice, procure a cookbook, follow its directions—and wind up with a cake as good as Betty Crocker's.

When it comes to automobiles, Chrysler Corp./Direct Connection has been thoroughly testing their performance recipes for years. Unlike other manufacturers, Chrysler tests the best of all available parts and pieces to come up with dyno-proven packages guaranteed to produce results when used as directed. They then put factory part numbers on these parts and carry them on their Direct Connection (DC) parts program, disseminating their availability and intended usage to the public through the automotive press and their own in-house publications. The result is a series of "go fast" recipes for Moparphiles, a veritable cookbook of combinations for producing a successful performance car.

Recently DC went back and re-evaluated their small-block "A" engine recipes, rechecking and modernizing their ingredients for the '80s. A complete battery of dyno research was conducted at Arrow Racing, where long-time Chrysler racers have extensive experience with this type of engine. In all tests, a basic 360 engine with stock block, stock cast crank, and stock-type pistons with deepened valve notches served as the dyno mule. Compression was kept at 9:1, so the local racer at-

PHASE 1
15-SECOND BRACKET—240 HP

1 stock cast-iron Thermoquad intake
1 stock Thermoquad 4-bbl. carb
1 set stock exhaust manifolds
1 stock distributor

1 DC chrome air cleaner, P4286575
1 360-2-bbl. hydraulic cam, 5214658
16 stock hydraulic lifters
2 stock 360 cylinder heads with stock springs, retainers, and keepers

Illustration: Gary McAllister

99

"A" MOTOR RECIPES

tempting to duplicate the packages could run on available premium pump gas (additional buildup details in Sidebar 1). The combinations tested using the mule are based on typical bracket-race e.t. "breaks"; in other words, the "15-second" package would conservatively enable the average 3300 to 3400-pound Mopar musclecar to turn low 15's in the quarter, assuming appropriate chassis mods—as described in the racing chassis manuals—were instituted at the same time (chassis book ordering information, see Sidebar).

PHASE ONE

How good is a stock 360? The baseline run was made using the stock-type components listed above, along with a DC chrome air cleaner. The stock 360 cam had .410 lift, 252 degrees advertised duration, 33 degrees overlap, and a 112-degree centerline. Two-degree initial timing was used, with 24 to 28 degrees total advance. The engine pulled from 2750 rpm all the way to its 240-hp peak at 4250 rpm. Torque peaked at 3750 rpm, where 315 ft.-lbs. was recorded (see Graphs). With this engine, a 3300 to 3400-pound car should be able to run in the 15's with no other mods.

High-volume oil pump requires a heavy-duty oil pump and distributor driveshaft.

DC's forged connecting rod is delivered ready to run, having already been shot-peened, magnafluxed, and fitted with high-strength nuts and bolts.

PHASE 2
14-SECOND BRACKET—291 HP

1 electronic ignition conversion package, P3690426*
1 set solid-core ignition wires, P4120716
1 .455-lift hydraulic camshaft and tappet package, P4286671
16 dual valvesprings, P4120249
16 aluminum valvespring retainers, P4286573 (optional)
Remove air cleaner
Max advance spark curve

* Includes (1) vacuum advance electronic distributor, P3690430, (1) "Orange Box" electronic ignition control unit designed for general purpose high-performance use, P4120505, (1) ballast resistor, P2095501, (1) control wiring harness kit, P3690152.

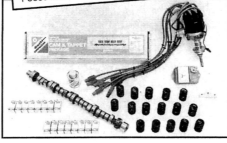

PHASE TWO

Phase 2 added DC hydraulic camshaft, P4286671. Introduced by DC as a replacement for the old, discontinued 340 high-perf cam, the new offering features .455 lift, 272 degrees advertised duration, 48 degrees overlap, and 112-degree centerlines. DC tappets and break-in lubricant are included in the package, but the recommended P4120249 valvesprings must be purchased individually. Stock valvespring

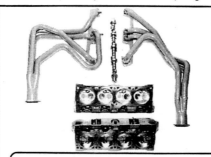

PHASE 3
13½-SECOND BRACKET—333 HP

1 set 1⅝ primary x 3-inch collector headers, P4286437
16 2.02-inch intake valves, P3690230
16 1.60-inch exhaust valves, P3690231 (optional)
1 Edelbrock Performer 2176 intake manifold, P4286531
1 .474-lift hydraulic camshaft and tappet package, P4286630
2 sets hardened valve stem locks (triple groove for new valves), P4120620
Perform bracket valve job

---

retainers can be used, although the DC aluminum high-perf pieces are preferred.

Electronic ignitions are highly recommended by Chrysler over obsolete breaker-point systems, since the electronic designs require less frequent tuneups, increase secondary voltage, improve starting, eliminate irregular timing and dwell changes, increase plug life, and eliminate point bounce—all of which increases rpm potential. The DC electronic ignition conversion package listed here is designed for those still stuck with the breaker point systems. If you have the Chrysler factory-electronic "smog"-type systems, the Orange Box is available for replacing the smog box in a kit that also includes lightweight advance springs (P4007968). The complete distributor comes with a quick curve already installed. Arrow installed the distributor using the "out of the box" curve. Total timing ended up at 35 degrees, with 10 degrees initial on the block.

With the new, still streetable camshaft and better ignition installed, the engine made more power throughout the rpm band. While the torque peak remained as before at 3750 rpm, it was up by 25 ft.-lbs., from 316 to 341 ft.-lbs. Overall, the cam and ignition tended to flatten and broaden the torque band, as is shown by the virtually flat peak between 3750 and 4000 rpm, as well as the shallow drop-off between 4100 and 5000.

Horsepower, too, was up throughout the powerband, with the gap widening at higher rpm. At the Phase 1, 4250-rpm horsepower peak, the new package made an additional 30 hp, and was still going strong. Horsepower now peaked at 5000 rpm, 750-rpm higher than before, where 291 hp was recorded.

This package will get the average 3400-pound car well into the 14's if 4.10 gears, slicks, and other mods as listed in the chassis book are also added.

PHASE THREE

At this point, it was time for some serious mods. Arrow Racing removed the heads and added DC's larger 2.02-inch intake valve. While not strictly necessary, they replaced the stock high-performance 1.60-inch exhaust with the DC 1.60-inch exhaust, since the latter is lighter and its head shape is designed for superior flow characteristics. A "bracket valve job," as described in the Chrysler racing manual, was performed. While the heads were off, a hotter DC hydraulic cam (P4286630) featuring .474 lift, 280 degrees advertised duration, 60 degrees overlap, and 110-degree centerlines was installed. This new cam supersedes the old Purple Shaft Street Hemi grind P3690213, which featured .471 lift and 284 degrees advertised duration. On paper, the new cam has a tad more lift and a little less duration, but the real key is the .050-lift duration

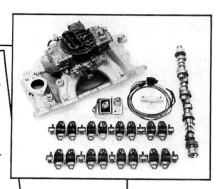

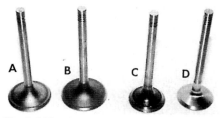

Stock 360 came with 1.88 intake/1.60 exhaust valves (A&C). Beginning with this test, they were replaced with DC heavy-duty intake and exhaust valves (B&D, respectively). DC intake has larger 2.02 head. DC exhaust's head, while same size as stocker, has better shape for improved flow.

numbers: 228 degrees for the Street Hemi, compared to 234 degrees for its replacement. Hence, the new cam opens and closes quicker, has more duration when you can really use it on the top end, yet offers improved idle and better torque. It works well with the DC headers listed, which fit virtually all Mopar passenger car chassis.

As can be seen from the graphs, these mods really woke the 360 up. Power was up both downstairs and upstairs. The torque peak was raised 500 rpm to 4250, where 378 ft.-lbs. was produced; 333 hp was recorded at the 5500 rpm peak, which was also 500 rpm higher than before. Comparing peak to peak, it comes out to 42 more hp and 47 more ft.-lbs. Also note how smooth the horsepower curve is between 4750 and 6000, not varying by over 10 hp. This definitely looks like the ultimate street cam for an A motor, and will get a 3400-pound car into the 13's with appropriate chassis mods and a heavy-duty torque converter or heavy-duty clutch.

PHASE FOUR
At this point we are starting to move radically away from streetability. Cam-

shaft P4120233 features .508 lift, 292 degrees advertised duration, 76 degrees overlap, and 108-degree centerlines. This is the biggest hydraulic offered by DC, and as can be expected from its radical specs, the cam has poor idle characteristics and a lack of bottom-end torque. DC says it's best suited for use with a manual trans and 4.30 rear axle ratio. Slightly milder cams are offered for automatic trans-equipped cars. Valve-to-piston clearance *must* be checked with this cam. Its radical nature mandates the use of heavy-duty hydraulic rocker arms and improved oiling "banana groove" rocker arm shafts. Adjustable pushrods may be required to retain correct valve-train geometry, especially if the heads have also been milled.

DC's new adjustable hydraulic pushrods offer an economical solution to valve-train geometry problems sometimes encountered when using the more radical high-lift hydraulic camshafts, highly milled heads, and/or reworked valves.

Once such a radical cam is installed, the Holley carb/manifold combo makes the most top-end horsepower. As delivered, the vacuum secondary 750-cfm carb is of "Model 4160" configuration, with a rear metering plate instead of a full secondary metering block with changeable jets; many competitors add on the rear block with an available aftermarket Holley kit.

The improved high-output DC electronic ignition "Chrome Box" is highly recommended for this increased horsepower and rpm application.

Not surprisingly, dyno testing showed considerable top-end increases, but compared to the other previous packages, the big cam/single-plane intake

Chrysler's New Cookbooks

Elsewhere in this article we continually refer to Chrysler's "racing manuals." Chrysler publishes more extensive hop-up information than any other automaker. While their racing books are fairly expensive at $15 each, the information is updated annually, and includes everything anyone would possibly want or need to know about setting up a Chrysler product for high-performance street or competitive racing applications. Chrysler's $4 high-perf parts catalog lists all the latest factory parts, plus covers total bracket race "combinations." No Mopar enthusiast should be without these publications. They can be ordered from any Chrysler-Plymouth or Dodge dealership, from your local Direct Connection speedshop, or by sending a check or money order (plus your state's sales tax) to: Direct Connection Catalog Center, 20026 Progress Dr., Dept. C85, Strongsville, OH 44136.

Current Chrysler Performance Publications

PART NO.	TITLE	SUBJECT
P4349340	**Engine Speed Secrets**	All Chrysler V8, 6-cylinder, and import 4-cylinder engines
P4349341	**Chassis Speed Secrets**	Rear-wheel-drive chassis and drivetrain prep, engine swaps, suspension mods, roll cage construction, rearends, transmissions
P4349222	**2.2/FWD Engine & Chassis Book**	Front-wheel-drive engine and chassis mods
P4286727	**Oval Track Modification**	Oval track chassis, drivetrain, suspension, brakes, body info
P4349086	**Performance Parts Catalog**	Up-to-date DC part numbers

"A" MOTOR RECIPES

"13 flat" package was down on horse-power below 4400 rpm, confirming its poor suitability for daily street use. Above 4400 rpm, it pulls rapidly away from the best previous package, and at 6000 rpm is still going strong—to the tune of 375 hp, a gain of 42 hp compared to the previous cam at its 5500-rpm peak.

The cam produces slightly less torque at low rpm than the 13.5 package. Its 373-ft.-lb. peak occurs at 4500 rpm, 250-rpm higher than before, where it is actually about 5 ft.-lbs. lower than the previous package's torque peak. You can see why it's recommended for a manual! But since the curve is virtually flat between 4250 and 5000 rpm, and the over 4500-rpm torque production is markedly superior, this cam not only would be good for manual bracket race cars, but also for limited sportsman oval track cars or highly modified gymkhana/Mulholland type activities.

PHASE 5
12-SECOND BRACKET—420 HP

2 A-engine ported bracket heads, P4286864*
1 set big tube headers (1⅞-inch primaries x 3½-inch collectors)**
1 mechanical camshaft and tappet package, P4120657
16 mechanical rocker arms, P3870827
1 rocker arm adjusting screw & nut package, P4120636
1 "Super Gold" electronic ignition control unit, P4120600
1 Accel race coil, P3690560
1 ballast resistor, P2444641
1 Holley 0-3418-1 850-cfm vacuum secondary carburetor, P3571012 (optional)

* Assembled heads include (8) 1.60-inch exhaust valves P3690231, (8) P3690230 2.02-inch intake valves, (16) P4007178 titanium valvespring retainers, (32) P4120620 hardened keepers, (16) P3412068 dual valvesprings, (16) P3690963 P-C valve seals. Head casting P4100409.
** Part numbers vary per chassis. See DC catalog.

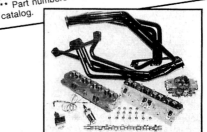

360 Short-block Buildup

Speed-Pro flat-top pistons were used. Valve relief notches were added to maintain adequate piston-to-valve clearance with the more radical high-lift camshafts tested.

All dyno results were tabulated by Arrow Racing's fully computerized Superflow dyno. The engine is assembled on a completely portable "rollaway" test stand, which makes engine changes easy.

The bottom end of the Arrow Racing 360 remains surprisingly mild. What mods were performed probably aren't necessary until at least the 13½-second bracket, although Arrow performed them initially to ensure the dyno mule's long-term survival. As an externally balanced engine, the 360 requires an externally balanced flywheel and harmonic balancer.

MAJOR INTERNAL COMPONENTS

BLOCK	Stock 1985 production, deburred and honed with torque plate
CRANK	Stock 1985 production, checked for straightness
PISTONS	Speed-Pro 2380P, with deepened valve reliefs for high-lift camshafts
RINGS	Speed-Pro R-9343+.005: 5/64 plasma-moly 1st, 5/64 cast-iron 2nd, 3/16 standard tension oil
MAIN BEARINGS	DC tri-metal P4286956 fully-grooved
ROD BEARINGS	DC tri-metal P3690683
CONNECTING RODS	DC P3690641 forged 8640 steel; delivered "magged" and shot-peened with high-strength nuts and bolts
HEAD GASKET	Fel-Pro 1009 (Fel-Pro supplied all gaskets for engine)
OIL PUMP	DC P4286589 high-volume
OIL PUMP DRIVE-SHAFT	DC P3690715, required with high-volume pump
OFFSET CAM KEY PACKAGE	DC P4286500, used to degree-in all cams tested "straight up"
DOUBLE ROLLER TIMING CHAIN	DC P4120262

Parts not listed are peculiar to each dyno package and are listed in the main article.

BUILD SHEET

For data not listed below, refer to stock service manual. All dimensions in inches or fractions thereof, unless otherwise stated. ENGINE BUILDER: ARROW RACING ENGINES, 224 South St., Rochester, MI 48063, (313) 652-0604.

DISPLACEMENT	360
BORE x STROKE	4.00 x 3.58
COMPRESSION RATIO	9:1
DECK HEIGHT	.025 below deck
CYLINDER HEAD VOLUME	68cc
COMPRESSED GASKET VOLUME	8.8cc
COMPRESSED GASKET THICKNESS	.039
PISTON-TO-VALVE CLEARANCE (0 lash)	
Automatic trans	.090 intake, .100 exhaust
Manual trans	.100 intake, .110 exhaust
PISTON-TO-HEAD (with gasket)	.055
ROD/MAIN BEARINGS	.0025
CRANK END-PLAY	.003-.007
PISTON-TO-WALL	.006
RING GAP	
1st	.012 minimum
2nd	.016 minimum
VALVESPRING INSTALLED HEIGHT	1.650 to 1.670
VALVE LASH (DC P4120657 mechanical)	.032 intake/exhaust
CAMSHAFT END-PLAY	.005*
HEAD BOLT TORQUE	105 ft.-lbs.
HONE PROCEDURE (Sunnen CK10)	Use 625 stone with torque plates installed on both cylinder heads (if only one plate is available, install cylinder head on opposite deck)

*Chrysler A-engine uses a camshaft retaining plate. This figure is the required cam-to-plate clearance.

PHASE FIVE

Direct Connection now sells a 340-type bracket ported A-motor head. Unlike the exotic W2 heads, it doesn't require special costly valvetrain pieces or manifolding. Specifically designed as a cost-effective solution for bracket racers, it comes ported, cc'ed, and ready to bolt on with racing valvesprings and titanium retainers already installed.

When combined with the newly designed P4120657 mechanical cam (.590-inch lift, 312 degrees advertised duration, 104 degrees overlap, and 106-degree centers), big-tube headers, and appropriate chassis mods, the average car should run in the low 12's. Adjustable rocker arms are required; use the "Econo W2" *exhaust* rockers

on both intake and exhaust positions. Retain the banana groove shafts from the previous package. Naturally, piston-to-valve clearance must be checked, and the ultimate Gold Box ignition system is highly recommended. Chrysler used to recommend an 850-cfm vacuum secondary carb for this package, but the new camshaft and recent tests confirm the required horsepower numbers with the more tractable 750. Incidentally, the big headers, when used on the Phase 4 combination, have been proven to be worth about 20 more horsepower. Of course, the bottom end suffers even more, although that could be somewhat alleviated by adding collector extensions.

Looking at the dyno trace, we see this combination is the weakest yet on

the bottom end. No way would it work on a street car; your opponent would be out of first gear by the time you got "up on the cam." "Up" in this case is about 4300 rpm. Above that point, the Phase 5 package literally races away from all previous combos. At 6250 rpm, 420 hp was produced, and the curve was still rising. As for torque, it was below previous packages until 4300 rpm; then it was "sayonara," with the 392 ft.-lbs.-peak occurring at 5000 rpm. If you're serious about cooking, this one's "microwave on high."

There are considerably quicker bracket packages than this (all the way down to the 8.80 Super Comp level), but beyond the 12-second break serious bottom-end mods (including lighter rods and high-compression pistons) become necessary.* Still, the fact that you can pick up 180 hp with a virtually stock bottom end and only 9:1 compression is rugged testimony to the Chrysler small-block basic stoutness. The important thing to remember is that these packages work because they are a coordinated combination—each should be installed in its entirety, or the actual performance gains will be disappointing. The 340-based motors should have similar results to the 360 dynoed in these tests, and some have the advantage of already starting with the big-valve heads. The 273/318 engines may require a slightly different package; check the appropriate racing manual.

Thanks to Direct Connection's considerable reasearch and up-to-date parts program, Chrysler's small-block is at least as cheap (if not cheaper) to build to a given performance level than the "most popular" small-block. And there's no guesswork involved—once you've attended Chrysler's cooking class. **HR**

*The 9.90 Pro Gas A-motor buildup is covered on page 64 of this book.

TEST RESULTS

All tests were conducted on Arrow Racing's computerized dyno, which corrected the results to standard sea-level atmospheric pressures and temperatures. Test numbers correspond to the text's numbered "recipe ideas."

————————	12-sec. pkg.
————————	13-sec. pkg.
————————	13.5-sec. pkg.
————————	14-sec. pkg.
————————	15-sec. pkg.

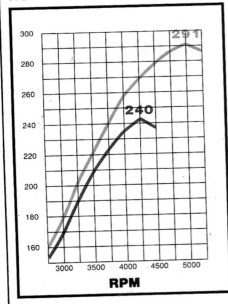

HORSEPOWER (hp.)

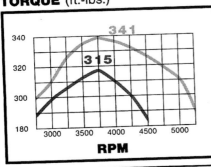

TORQUE (ft.-lbs.)

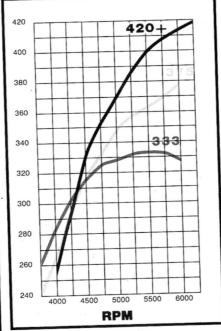

HORSEPOWER (hp.)

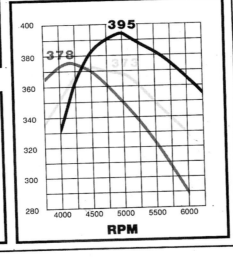

TORQUE (ft.-lbs.)

TITLE BOUT
THE REMATCH
MORE POWER, MO' POWER, MOPAR!

A MOPAR JOINS THE CUBIC-INCH CLASH

By Will Handzel

As we outlined last month, HOT ROD is conducting a shootout between Chevrolet, Ford and Chrysler big-block engines in an altered version of the "Title Bout" story that we did back in the October '92 issue.

That first Title Bout was a crate-motor small-block shootout with only a few rules. This time around, there were just six engine rules laid down—engine displacement must be between 480 to 520 ci, engines must run on pump gas (Unocal unleaded 92), induction is limited to a single four-barrel carb on a cast intake, no nitrous or superchargers allowed, no symmetrical-port heads permitted, and the engine must run through mufflers. No engine rules other than that—a true run-what-you-brung competition.

As was the case with the original Title Bout, these engines will be compared on the dyno and also in cars. To make the comparison valid, all three cars were outfitted with 4.11:1 gears, a three-speed automatic trans without a trans brake, a 10-inch torque converter, 29x10 Goodyear slicks and a stock-type rear suspension for the make of the car.

Next month, the Chevy big-block goes together, and the big guns play it out on the black and sticky quarter-mile at Bakersfield Raceway in the July issue. It's gonna be good, so don't miss it!

HERB McCANDLESS PERFORMANCE BUILDS A 736HP PUMP GAS 500CI CHRYSLER

OK, kids. Hang on to your hats 'cause this baby is gonna make some noise! As the second of the three entrants in HOT ROD's big-block shootout, the Mopar camp is coming out of the gate with both barrels ablaze. If you're a bettin' man, then get your money down, 'cause when you bring Herb McCandless, his able engine builder Ken Lazzeri and his other fast friends to any party, it's gonna get exciting. McCandless has been involved in drag racing since the '60s, when he match-raced his Chryslers all over the country against the likes of Dyno Don Nicholson and Bill "Grumpy" Jenkins, so he has his own ideas on how to get down the strip quickly.

The approach that the McCandless troops chose when building this engine might surprise you, but they spent a lot of time testing different combinations and feel they have fig-

The McCandless formula for 500 cubic inches of pump gas power includes a cast-iron 400ci block, aluminum Indy Cylinder Heads 13-degree heads and intake, a Mopar Performance 4.150-inch steel crank, Wiseco pistons, BME 426 forged-aluminum rods, a Cam Motion bumpstick activating Jesel and Competition Cams valvetrain pieces, a BG carb, MSD ignition, a BHJ balancer and an oil pan by Charlie Gilbertson. The parts have been massaged to make the package work—and that's where the story is.

Chrysler big-blocks use the center main bearing as the thrust bearing surface, unlike a Chevy, which uses the rear main bearing for this purpose. An ARP main cap stud kit was used along with C&A Assembly lube on all C&A bearing surfaces. Notice the narrow oil pan rail after the clearancing.

Since this engine wasn't going to make the same power as the high-compression race engines, it was felt that cutting away the oil pan rails (*top*) and bolting on a special wide pan from Charlie Gilbertson would reduce oil windage and parasitic power losses without increasing the chances of the block failing.

Also to reduce parasitic losses, the Mopar Performance crank was sent out to Winberg Cranks, to be lightened (by 9 pounds) and aerodynamically shaped.

The 6.570-inch BME aluminum rods are actually old fueler rods that can take up to 600 runs or be run on the street with the durability of a steel rod, according to Lazzeri and McCandless. The Wiseco piston, C&A pin and rod had a total bob weight of 1900 grams.

ured out how to run fast within the rules. Says Lazzeri, "The rules required us to think through what we wanted from the engine. The Unocal 92 unleaded really was a wild card, because we usually work with race gas. The 92 is good gas, but we knew it just wouldn't take the compression, rpm or timing we usually put to our race gas combinations. So we decided to run a low-compression engine, 9.5:1, that wouldn't make a lot of peak horsepower but would make peak torque around 5200 rpm. That is the average rpm we determined the engine would operate in during a run, based on the transmission, rearend gear and tire rules."

Instead of starting with a 440ci block, a 400ci block was chosen because it is 35 pounds lighter and is considered stronger than the 440 block. The 400 block requires a shorter rod than can be used in the 440 block, but it was felt that the short rod would help provide a flat torque curve. All machine work was performed in the McCandless shop. The block was cut for O-rings because copper head gaskets were used to dial in the compression—Lazzeri could vary the compression from 8.4 to 10.4:1 with just the gaskets.

By using a lightweight, 4.375-inch-diameter Wiseco piston that locates the pin as high as possible on the piston, a 6.570-inch rod was used with a 4.150-inch-stroke crank to net 499 cubic inches. Each flattop piston weighs 480 grams after the valve reliefs were cut

The ring combination on the Wiseco pistons includes Childs & Albert Duramoly .031-inch top and second rings and a low-tension oil ring pack. Lazzeri uses a Dykes top ring and a backcut second ring to seal the combustion chamber.

With the 92-octane unleaded and low compression, the engine could develop a tendency to act lazy in revving under load, so one of the goals was to reduce the parasitic losses and keep the rotating weight to a minimum. To help keep as much oil off the crank as possible, the oil pan rails were ground heavily, and a special, wider oil pan with a drop windage tray was used. Also, the crankshaft was aerodynamically shaped and lightened by Winberg Cranks to allow it to cut through the air more efficiently, again reducing the power losses. The valvetrain was likewise scienced out to eliminate as much flex and misalignment as possible.

Aluminum rods and main caps are used to reduce the shock to the 400 block, which typically fails around the main webbing in high-horsepower applications. The aluminum absorbs some of the shock, allowing the block to withstand the stresses. Also, the block is half-filled with Hard Block for added strength.

While you might think aluminum rods were used to reduce weight, you'll be surprised to know the BME aluminum rods weigh the same as the stock steel rods. McCandless uses the aluminum rods because he feels they reduce the shock to the block. This is also the reason aluminum main caps are used. To solve the strength problem with these blocks, Indy Cylinder Head is just putting the finishing touches on a completely redesigned engine block that has enough extra

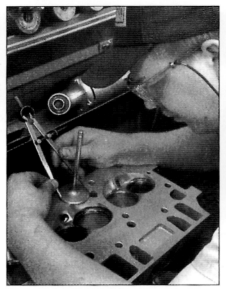

The 13-degree aluminum Indy heads are a new version of the successful big-block Chrysler head. The heads McCandless used were machined (on Indy's CNC mill) with the valve stem centerline at 13 degrees off the cylinder bore centerline, and the valve centerline was moved closer to the centerline of the cylinder bore to reduce the valve shrouding. The intake valves are 2.19, and the exhaust valves are 1.81 inches in diameter.

A standard Indy race head is on the bottom while the set used on this engine is on top. Notice how much smaller the pump gas intake ports are. Indy Cylinder Head kept filling the ports until the flow bench numbers started to fall off with the 2.19/1.81-inch valve sizes, to end up at this size.

material to be punched out to 600 ci and handle over 1000 hp.

The unfair advantage that the Chrysler camp has in its favor is the Indy heads. They are well-designed out of the box, and the pair used for this engine was optimized for this application through plenty of flow bench and

SOURCES

Barry Grant Fuel Systems
Dept. HR05
Route 1 Box 1900
Dahlonega, GA 30533
706/864-8544

B-H-J Products, Inc.
Dept. HR05
37350 Enterprise Ct., #2
Newark, CA 94560
510/797-6780

Bill Miller Engineering (BME)
Dept. HR05
4895 Convair Dr.
Carson City, NV 89706-0492
702/887-1299

Cam Motion
Dept. HR05
2092 Dallas Dr.
Baton Rouge, LA 70806
504/926-6110

Charlie Gilbertson Oil Pans
Dept. HR05
5281 S. Hametown Rd.
Norton, OH 44203
216/825-3586

Childs & Albert
Dept. HR05
24849 Anza Dr.
Valencia, CA 91355
805/295-1900

Competition Cams
Dept. HR05
3406 Democrat Rd.
Memphis, TN 38118
901/795-2400

Herb McCandless Performance Parts
Dept. HR05
P.O. Box 741
Graham, NC 27253-0741
910/578-3682

Indy Cylinder Head
Dept. HR05
8621 Southeastern Ave.
Indianapolis, IN 46239
317/862-3724

Jesel
Dept. HR05
1985 Cedarbridge Ave.
Lakewood, NJ 08701
908/901-1800

MSD Ignition
Dept. HR05
1490 Henry Brennan Dr.
El Paso, TX 79936
915/857-5200

Winberg Crankshaft
Dept. HR05
2200 S. Jason
Denver, CO 80223
303/922-0741

WISECO Piston, Inc.
Dept. HR05
7201 Industrial Park Blvd.
Mentor, OH 44060
216/951-6600

Lazzeri and McCandless thank Kip Fabrea and Rick Womble at Cam Motion for helping to design a camshaft that would make power from 3000 to 7000 rpm without sending the engine into preignition or detonation from high cylinder pressures. The dual-pattern cam was ground to 276/281 degrees at .050 for the intake and exhaust, with .782 /.748 max lift.

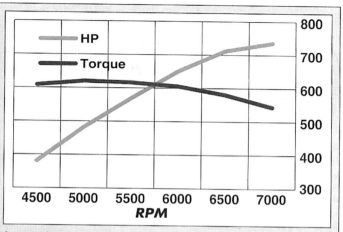

The engine was bolted onto the dyno, brought up to operating temperature with the MSD distributor set at 32 degrees before TDC, and the hammer was dropped. While this engine was designed to make peak power at 6800 rpm, the power actually peaked at 736 hp at 6900 rpm. Max torque of 626 lbs-ft came at 5100 rpm. More importantly, check out the flat torque curve between 3000 and 7000 rpm, which will propel this car down the quarter.

The pistons sit .140 inch down in the hole at TDC. With a combustion chamber volume of 65 cc, the engine has a 9.5:1 compression ratio—well within the limits of the Unocal 92 unleaded.

dyno time. The intake and exhaust ports were reduced in size, from 370 to 350 cc, to maintain port velocity with 9.5:1 compression; the valve angle was reduced two degrees, to 13 degrees total, by angle-milling; and the valve centerlines were moved closer together and centered in the bore, to minimize valve shrouding against the cylinder walls. These modifications have shown power gains in many forms of racing in the last few years (most notably, the Yates-modified Ford head in Winston Cup), so it should come as no surprise that this engine made a lot of power

with these changes.

While the cylinder heads are critical, this type of engine was relatively new to the McCandless boys, so they needed some help designing the roller cam. Since cylinder filling is difficult with a lower-compression engine because there is less vacuum as the piston goes down in the bore (just as there's less compression as the piston rises), the cam needs to really be dialed in to fill the cylinder completely and make power at higher rpm. For help, the crew called upon Kip Fabrea and Rick Womble at Cam

An adjustable Jesel belt drive was bolted on with the cam advanced four degrees.

SUPPLIERS	PART	PART NO.
Indy Cylinder Head	heads	440-1-13
	valve covers	440-7
	intake	400-2
MSD (Mopar Performance part No.)	distributor	MP P4120942
	bronze gear	P3690876
Mopar Performance	crank	P5249338
	oil pump	P4286590
	cam bearings	MP4120261
Jesel	timing belt	35000
	rocker arms	7238
Wiseco	pistons	499-1
Bill Miller Engineering	rods	MCCBME
Fel-Pro	valve cover gaskets	1610
	intake gaskets	MCCMAX
Cam Motion	cam	CMI280RLR
Competition Cams	pushrods	7918CC
	valve seals	503-16
	retainers	733-16
	locks	611-16
K-Motion	valvesprings	K1000
B-H-J	harmonic balancer	MPES7
Childs & Albert	rings	SPRR10306-.005
	pins	.990x2.930
	rod bearings	R4262STD
	main bearings	M3830STD
BG Fuel Systems	carburetor	93755

Motion to put together a cam profile that would work. After some dyno testing with various cams, the cam profile they settled on uses an increased exhaust centerline, which

Competition Cams makes a set of roller lifters that has a Chevy cup offset .180 inch, so a Chevy pushrod can be used. These lifters eliminate the flexing and failures that have plagued these engines in high-horsepower and -rpm applications.

Jesel 1.7:1 shaft-mount rockers make the factory setup obsolete by reducing the chance of flex or failure, and when combined with the Comp Cams pieces, they make the valvetrain bulletproof. K-Motion valvesprings were combined with Comp Cams retainers and locks to control valve motion.

A swinging pickup from Milodon was modified to sweep the bottom of the oil pan. The tube in the pan is for the center link of the steering linkage to pass through. The advantage of the Chrysler is that the wet-sump oil pump has external pickup lines, allowing the pan to be scavenged wherever needed for optimal oil control.

Cam Motion felt allowed the engine to run and accelerate much more smoothly through the rpm band. The intake has a maximum lift of .782 inch, while the exhaust has .748 inch max lift. The duration at .050 is 276 and 281 degrees for the intake and the exhaust, respectively. The lobe centerlines are 106 for the intake and 114 degrees for the exhaust. The cam was installed at 106 degrees.

By combining the best valvetrain pieces from Jesel and Competition Cams, the misalignment and flex in the system were eliminated, which ensures the cam timing is actually being carried out on the valve, and practically eliminates the chance of failure.

With the engine bolted together and placed on the dyno with 2-inch Hooker Headers, some pulls were made with the MSD ignition set at various timing settings. As the combination was dialed in, it was discovered that the engine did not like any timing over 32 degrees

This photo should give you an idea of how wide the oil pan is. The bolts to hold the pan on the block need to be accessed through plugs in the bottom of the pan!

The Indy valve covers have an intricate oiling system installed to cool and lubricate the valvesprings. McCandless constructed this to help the valvetrain last. This system is plumbed from one of the oil galleries in the back of the block.

BTDC—beyond that the engine exhibited detonation and preignition—so the timing was set at 32 degrees for all testing.

During some initial testing with a mule motor, Lazzeri noticed the engine would labor to rev during the step tests, but the final combination seemed to run strong through an entire pull, which he felt was the result of all the work done to refine the engine to run on the Unocal 92.

McCandless and Lazzeri took the usable power with a lightweight approach to this engine package, instead of going for killer peak numbers. Will it pay off? Can the smallest cubic-inch engine in this matchup play with the big dogs? Stay tuned as this brawl is settled once and for all. **HR**

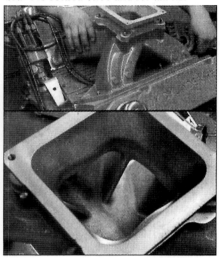

The Indy intake needed to have its runners filled the same amount as the cylinder heads for the best flow numbers. Dividers were extended and blended into the plenum area, and the spacer was also blended into the intake. The MSD ignition provides a rock-solid spark for the engine.

The 9375 Holley was seriously breathed on by BG Fuel Systems. The trick BG fuel regulator mount was fabbed by McCandless.

HALF PRICE HEMI

By Steve Magnante

Photography: Steve Magnante

ver since its introduction in 1964, the Chrysler 426 Hemi has gained a reputation as one of the most potent engine designs ever released to the general public. Its only real drawback is expense. Hemi parts, new or used, are among the most costly in the automotive king-dom. But what if you could substitute a dirt-cheap 440 wedge block for the big-buck Hemi? You'd save thousands. Stage V Engineering has been making Hemi power more affordable since 1987 with its cylinder head conversion. Starting from scratch, Stage V Engineering proprietor Eric Hansen constructed custom tooling which produces a high-quality cast-aluminum replica of the 426 head that incorporates specific changes which allow it to bolt directly atop a lightly modified wedge block. We caught up with Dale Reed as he constructed a 9.8:1 520-inch Stage V Hemi Conversion motor. Though the extra equipment involved brought the cost to about $12,000, we'll outline how a budget version, using mostly stock internals, can be assembled for about $7,500.

How to Turn a Wedge Into an Elephant on the Cheap

Half-Price Hemi

We purchased a '69 Chrysler Newport 440 block for $80 at a self-serve auto wrecking yard. Though it lacks the cross-bolted main bearing caps of a 426 Hemi block, the RB-series 440 wedge (and its 413-inch siblings), can reliably maintain 900 hp with only minor modifications. It weighs 213 pounds, making it about 70 pounds lighter than a Hemi block. When coupled with the aluminum Stage V Conversion heads, the finished engine will weigh around 580 pounds (140 less than a production Street Hemi). Any 413 or 440 block that passes pressure and magnaflux tests is acceptable, although blocks cast after 1975 are thinner and must be sonic-checked prior to overboring. The B-series low-deck 361, 383, and 400 Chrysler big-blocks can also be used as the basis for a Stage V Conversion; commercially available intake manifolds are nonexistent, so fabrication would be necessary.

With more than 650 hp on tap, we added a measure of reliability by replacing the stock ½-inch main cap bolts with ARP studs that offer a 60 percent increase in clamping ability. The three center main caps and bearing bulkheads were drilled and tapped to accept ⁹⁄₁₆-inch fasteners that are tightened to 130 lb-ft; the two outboard positions retain stock-size fastener diameters, and are torqued to 100 lb-ft. All casting flash was removed, and the main oil feed holes were enlarged to ⁹⁄₃₂-inch on 2 and 3, and ⁵⁄₁₆-inch on 4 to increase the volume of flow.

Because the longer stroke increases the arc of the connecting rods, the stock oil pickup tube must usually be replaced by an expensive external oiling system. But thanks to our reworked 2.20-inch rod journals, we were able to employ connecting rods with smaller big-end dimensions—the only area that requires grinding is shown here. Offset drilling the stock ⅜-inch diameter oil pickup tube hole to ½-inch moves the oil pickup tube outboard and away from the connecting rod, while increasing oil flow capacity. Thanks to these procedures, we were able to retain an economical OE-type oiling system using a Milodon 7-quart deep sump pan (PN 31010), Milodon high-volume pump (PN 18790), and Milodon extended ½-inch pickup tube (PN 18330). The machinations of the stroker precluded a windage tray. Many of the top Super Stock Hemi builders have summarily abandoned this item, claiming it does as good a job of keeping oil away from the crank as it does keeping oil from returning to the pan quickly.

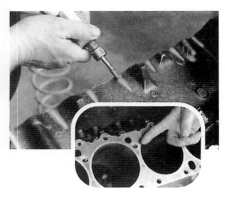

Although the Conversion heads bolt on, the wedge block requires pushrod reliefs in the walls of the lifter valley. The notches are ⅛-inch deep and can be done by hand with a die-grinder and carbide burr—Stage V can supply a template for this operation. To ensure maximum gasket sealing, enlarge the center row of head bolts from ⁷⁄₁₆- to ½-inch as per the instructions in the Stage V kit. We got our block ready on the garage floor in less than two hours. Gene Ohly at Evans Speed in South El Monte, California, (626/444-2838), gave it a 0.030-inch over-bore and honed it with a torque plate attached.

Above right: Since we saved so much dough on the wedge block, we decided to convert some of it to serious cubic inches—520 of the little buggers to be exact. We scored a used Keith Black 4340 steel billet stroker crank from a Top Alcohol Funny Car racer for $500, then had it prepped by Keith Black Racing (562/869-1518). It was checked for cracks, the main journals went 0.010 under, and the rod journals were offset-ground to decrease the stroke from ¾ to ⅝ (for a total of 4.375-inches). Reducing the rod journal diameters to Chevy big-block size of 2.200 inches (stock Chrysler is 2.375) greatly increases clearance inside the block. Finished, the billet piece weighs 75 pounds (a stock-stroke forged Hemi crank weighs 71). When you get your used stroker, make sure there is enough material in the rod journals to withstand the offset-grinding required for stroke reduction. Some of these cranks have rifle-drilled crank pins which will not tolerate offset grinding because it makes the rod journal too thin. Have your crank shop inspect the piece before you buy.

Half-Price Hemi

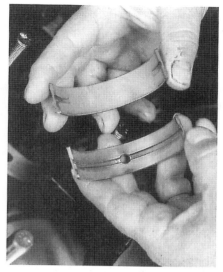

The wedge block does not have suitable oil return holes in the deck, so the Conversion heads must employ external oil return lines. This requires drilling and tapping two ⅜-inch pipe thread holes in the block. Dual-purpose buildups like this one are fine with a single return line running from the back of each head to the side of the block. All-out applications should adopt oil returns at the front of each head as well to prevent the valve covers from filling with oil faster than it can drain back into the pan. The braided stainless steel return line kit with the AN fittings shown here is available from Stage V for $165; you could fabricate your own for less.

Though some will argue, we've found that Arias forgings are the best for a street-driven Stage V Conversion motor. Due to their high silicon content, they experience minimal expansion at operating temperature. Consequently, the engine builder can reduce piston-to-wall clearance to create a tight motor that doesn't burn oil. Reed set ours with 0.006-inch skirt clearance. The standard-bore 440 wedge cylinder measures 4.320 inches, which is 0.070-inch larger than a standard 426 Hemi bore. A block overbored 0.030 or 0.060 would require a piston with a diameter larger than Chrysler ever offered over the counter for the Hemi. For pump gas compatibility, Dale selected Arias 291-828-01 slugs that feature a nominal 9.8:1 compression ratio. Even with our high-lift camshaft, generous relief notches provide in excess of 0.150-inch valve-to-piston clearance. These custom Arias forgings check in at 724 grams; that's less than a forged 440 Six Pack flat-top replacement piston! Gene Ohly balanced the assembly; thanks to the low mass of the parts, the bob weight total was 2,471 grams. Compare that to 3,095 for a stock 426 Street Hemi, and you can see that our 520 will maintain excellent throttle response and long rod life.

Though we splurged on the ex-Top Alcohol stroker crank, any '66-'74 forged steel 440 crank will perform reliably up to 800 hp in a stock stroke buildup. Later ('74-up) 440s had cheap cast-iron cranks worthy of 500 hp. When it comes to connecting rods, play it safe. Crower Lightweight Billet rods (PN 93911B) are milled from 4340 steel, are approximately three times stronger than a stock wedge rod, and weigh 859 grams (a forged Street Hemi rod weighs 1,068 grams). Custom-domed pistons fill the 169.5cc hemispherical combustion chambers. The port layout of a Hemi head is different than a wedge (I-E-I-E-I-E-I-E versus E-I-I-E-E-I-I-E) requiring a Hemi-specific camshaft, which fits right in. Dozens of off-the-shelf flat-tappet cams are available, but engine builder Reed heartily recommends the Crower No. 33471, a billet steel roller with 0.626 lift (I), 0.631 lift (E) and 298 degrees (I), and 304 degrees (E) advertised duration for acceptable street manners and banzai power output. Crower solid roller lifters (PN S-66232) work in concert with hollow Crower pushrods (PN 70399) to provide additional oil to the rocker arm adjusters at low engine speeds—right where stock Hemis tend to wear rapidly and require frequent valve lash adjustment.

Reed shuns fully grooved main bearings in the lower position because the grooves are 0.200 wide and result in a 20-percent reduction in surface area (from 0.900 to 0.700). He uses grooved inserts in the block and puts non-grooved inserts in the main caps. Though this is foreign to traditional wedge and Hemi thinking, Dale Reed has 30 years experience with Elephants, so we listened. The hard-face main bearings are heavy-duty Federal Mogul shells (PN 119M, 0.010 undersize) clearanced at 0.003 inch. The Federal Mogul rod bearings (PN 8-7200CH) are essentially big-block Chevy items that accommodate the reduced rod journal diameters of the crank and the Crower rods. Smaller rod bearings operate with less friction and demand less oil than stock-sized counterparts. This is the same setup used in blown alcohol cars (Pro Stockers rely on small-block Chevy 2.10-inch rod journals!). Rod bearing clearance is set at 0.002, side clearance is 0.015, and crankshaft endplay is 0.009-inch.

Ring packs are Speed Pro (PN R-9278/35), 1/16 chrome-moly top (0.018-inch), 1/16 iron second (0.014), and 3/16 oil scraper. End gaps are positioned 90 degrees apart during assembly. Note that our 440 block has round steam holes (A); in 1973, Chrysler enlarged them to a figure-eight shape. The later blocks are not as desirable in race buildups because these larger passages weaken the deck adjacent to the head-bolt holes. When the head bolts are torqued, they actually pull this area up and distort the cylinder wall approximately 0.0005 inch. Such flexibility has a natural effect on ring seal because the bores are no longer perfectly round. Despite this, the figure-eight block remains a viable candidate for general high-performance use, as the bore distortion is of little consequence to all but the most serious buildups. The pushrod relief notches mentioned earlier are clearly visible here (B).

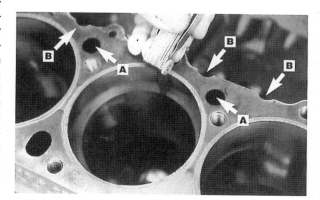

Half-Price Hemi

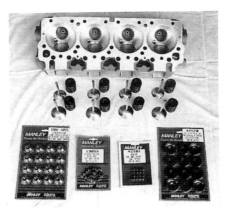

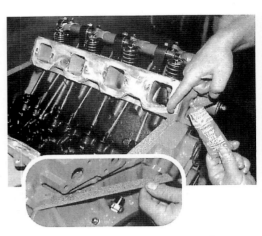

The Crower roller cam is linked to the crankshaft via a Cloyes Tru-Roller timing set. Slack in our stock timing set indicated that aggressive machine work during a previous rebuild had reduced the distance between the crank and camshaft. Reed specified a custom-fit Cloyes "problem solver" timing set (PN C-134) with shorter links. An ARP cam bolt kit (PN 144-1001) secured the top gear to the cam. The lobes and lifters of a flat-tappet cam are designed to gently push the camshaft toward the rear of the block, but a roller cam and lifters lack this capability, so a thrust button (Crower PN 86091) is required. Actually a miniature Torrington needle bearing assembly, the thrust button settles against the inside of the timing cover to limit the camshaft movement in the block.

With the optional street and strip porting package ($950), our heads are capable of flowing 414 cfm (intake) and 317 cfm (exhaust) at 0.700 lift, figures that surpass those of a typical vintage Street Hemi head. All Stage V Conversion heads use high-nickel chrome intake and heat-treated chrome-moly exhaust insert seats along with stainless exhaust and chrome-moly intake guides with K-line bronze inserts. We fleshed out our heads with Manley components: Severe Duty stainless valves (PN 11516, intake) and (PN 11901, exhaust), chrome silicon double-valve springs with dampers (PN 22441), spring cups to protect the aluminum heads (PN 42122), 10-degree locks (PN 13093), lash caps (PN 42101), and titanium retainers (PN 23661). The 2.20 intake and 1.90 exhaust valves are the same size as those in '65 A990 Race Hemi aluminum heads.

The Conversion heads accept any stock-type (16-bolt) Hemi intake manifold. To quell leakage, Reed uses brush-on liquid gasket adhesive to fix the intake gaskets to the heads, then coats the exposed surface with white grease. This provides a good seal while allowing easy removal and increasing the chances that the gaskets can be reused. Stock end seals, however, are only 0.516-inch thick. Reed offers custom end seal kits that feature 0.185-inch rubber impregnated cork gasket material. He glues the ½-inch wide strips (*inset*) to the block with liquid gasket maker, then applies a small dab of RTV sealant to the end joints, and finishes by smearing a very thin film of it over the cork.

Reed puts Conversion head to wedge block, but there's no need to call the chiropractor. Cast from virgin aerospace-quality C-355 aluminum alloy (22 percent stronger in yield than the 356 alloy commonly used in aftermarket castings), Conversion heads weigh only 38 pounds per side assembled—a similarly equipped iron Street Hemi head weighs 64 pounds. Exquisitely crafted roller rocker arms are investment cast from 17-4 stainless steel. All but three of the intake rockers are specific to Conversion motors and cannot be used on production Hemis. Despite apparent complexity, the Conversion valvetrain is stable to well over 8,000 rpm.

To isolate the exhaust pushrods, each intake port utilizes ⅝-inch-diameter press fit tubes. Despite the intrusion, the impact on port flow is minimal, and leakage is not a problem. Likewise for the countersunk attaching bolt in each intake port floor (*arrow*). Hemi blocks have cast-in bosses which engage studs protruding from deck surface of Hemi heads. Wedge blocks lack this provision, so Stage V Conversion heads simply use the stock wedge upper head bolt holes and port floor ⁷⁄₁₆ fasteners (with Nyloks) torqued to 55 lb-ft.

For our street-and-strip combo, we used a vintage Street Hemi intake manifold and Edelbrock 750-cfm manual-choke carburetors (PN 1407) which are nearly ideal out of the box. Because these carbs feature huge 1¹¹⁄₁₆ throttle bores all the way around, Joe Jill at Superior Automotive in Garden Grove, California, (714/894-9711) removed the plenum dividers for the necessary clearance. Similar to the heralded "Arlen Vanke" modification so popular among A/Stock Hemi racers in the '60s, this increases plenum volume and boosts top end breathing characteristics.

The Elephant Man

When you find an engine builder who doesn't advertise, you've got a good one. Such is the case with Dale Reed, proprietor of Dale's Place in Baldwin Park, California. Word of mouth is all he needs to keep busy. Beginning with an aluminum Max Wedge Plymouth that he bought new in 1964, then moving through a string of Hemi-powered Super and Pro Stockers, Reed has been a prominent force in the SoCal Mopar drag race and restoration scene ever since. Dale's Place specializes in meticulous, high-quality Max Wedge and Hemi engine assembly work. We are grateful for his assistance.

KING KONG DELUXE = $11,925

Displacement: 520ci
Horsepower: 650 at 5,460 rpm
Torque: 625 lb-ft at 4,600 rpm
Compression Ratio: 9.8:1
Bore/Stroke: 4.350x4.375
Bore/Stroke Ratio: 0.994:1
Rod/Stroke Ratio: 1.55:1
Maximum Safe Engine Speed: 7,500 rpm
Recommended Shift Point: 6,500 rpm

Dale Reed and his baby

Bottom End (subtotal = $3,203)
Block: '69 Chrysler 440 ($80)
Crank: Keith Black billet ⅜-inch stroker, used ($500)
Crankshaft Machining: Keith Black ($250)
Balancer: Mopar Performance, stock type PN 3830183 ($217)
Rods: Crower 4340 billet, 6.770 C-to-C, 2.200 big end dia. ($1,133)
Pistons: Arias forged high silicon, 4.350 bore ($752)
Piston Pins: Arias taper wall 0.990x2.75-in. ($80)
Rings: Speed Pro, chrome moly ($120)
Rod Bolts: Crower (incl. w/rods)
Main Cap Fasteners: ARP 9/16 main stud kit PN 140-5401 ($71)

Oiling System (subtotal = $420)
Oil Pan: Milodon 7 quart ($145)
Oil Pump: Milodon high volume ($75)
Oil Pickup: Milodon, ½-inch id ($35)
Oil Drain Line Kit: Stage V Engineering ($165 set)

Heads (subtotal = $6,398.50)
Type: Stage V Engineering hemi Conversion ($1,695)
Porting: Optional Stage V Engineering street-and-strip port job. At 0.700 lift, intake flow: 414 cfm (with floor bolt in place), exhaust flow: 317 cfm ($950)
Valvesprings: Crower double w/dampener ($114)
Valvespring Cups: Manley ($46)
Valves: Manley Severe Duty stainless, 2.20 intake, 1.90 exhaust ($410)
Retainers: Manley titanium ($136)
Locks: Manley, 10-degree ($37)
Lash Caps: Manley ($28)
Valve Seals: Pioneer neoprene with teflon insert ($16)
Rocker Assembly: Stage V Engineering, incl. roller rockers, thrust clamps, stands, shafts and plugs, (less standard adjusters and nuts) ($1,955)
Optional Polish on Rocker Arms: ($48)
Adjusters: Manton ⅜ high alloy with nuts ($224)
Pushrods: Crower ⅜ chrome moly, hollow point for roller lifters ($140)
Valve Covers: Stage V cast aluminum black powdercoated ($350)
Valve Cover Gaskets: Stage V Engineering ($12.50/set)
Head Gaskets: Stage V Engineering/Fel Pro ($34/set)
Head Bolt Kit: Stage V Grade 9 ($145)
Spark Plug Tubes: Mopar Performance PN P4120294 ($58)

Camshaft (subtotal = $708)
Type: Crower mechanical roller ($263)
Advertised Duration: 298° intake, 304° exhaust
Lift: 0.626-inch intake, 0.631-inch exhaust
Centerline: 106°
Lifters: Crower mechanical roller, oiling type ($385)
Timing Chain: Cloyes double roller ($60)

Induction (subtotal = $630.50)
Intake Manifold: '68 Street Hemi, used, plenum dividers removed ($200)
Intake Gaskets: Stage V Engineering ($12.50 set)
Carburetion: Dual Edelbrock 750-cfm, manual chokes removed ($418 pair)

Ignition (subtotal = $565)
Distributor: MSD ProBillet PN 8546 ($240)
Amplifier: MSD 6AL PN 6420 ($171)
Coil: MSD Blaster 2 PN 8202 ($29)
Wires: MSD HeliCore, 8mm for Hemi PN 31289 ($125)
Timing Specs: 34 deg. BTDC
Vacuum Advance: none

KING KONG ON A BUDGET = $7,511.35

Displacement: 440ci
Horsepower: projected 500 at 5,300 rpm
Torque: projected 500 at 4,300 rpm
Compression Ratio: 10.0:1
Bore/Stroke: 4.320x3.750
Bore/Stroke Ratio: 1.152:1
Rod/Stroke Ratio: 1.80:1
Maximum Safe Engine Speed: 6,000 rpm limited by stock rods
Recommended Shift Point: 5,500 rpm

Bottom End (subtotal = $1,269.40)
Block: Rebuildable Chrysler 440 short-block ($250)
Crank: Stock Chrysler 440, forged is preferable. Used (included w/short-block)
Crankshaft Machining: Local machine shop ($75)
Balancer: Used big-block (included w/short-block)
Rods: Used 440 (included w/short-block)
Pistons: Arias forged high silicon 4.32 bore ($752)
Piston Pins: Arias 1.094 dia. ($62.40)
Rings: Speed Pro OEM-type pre-gapped moly ($65)
Rod Bolts: ARP ⅜-inch PN 145-6002 ($65)
Main Cap Fasteners: Stock ½-inch bolts (incl. w/short-block)

Oiling System (subtotal = $85)
Oil Pan: Used stock (incl. w/short-block)
Oil Pump: Mopar Performance high-volume PN P4286590 ($55)
Oil Pickup: Used stock (incl. w/short-block)
Oil Drain Line Kit: Owner-fabricated, petroleum-certified flexible hose, brass ends ($30.00)

Heads (subtotal = $5,048)
Type: Stage V Engineering hemi Conversion ($1,695)
Porting: Base level Stage V Engineering bowl and port blend with seats cut and 3-angle valve job, ready to run. At 0.700 lift, intake flow: 330-cfm (with floor bolt in place), exhaust flow: 250 cfm ($260)
Valve Springs: Manley, single w/dampener PN 22406 ($67)
Valve Spring Cups: Manley ($46)
Valves: Mopar Performance 2.25 intake, 1.94 exhaust ($210)
Retainers: Manley chrome-moly PN 23635 ($48)
Locks: Manley 10-degree machined PN 13093 ($37)
Lash Caps: Manley ($28)
Valve Seals: Pioneer neoprene with Teflon insert ($16)
Rocker Assembly: Stage V Engineering, incl. roller rockers, thrust clamps, stands, shafts, and plugs ($1,995)
Adjusters: Stage V Engineering, ⅜ standard duty with nuts (included with rocker assembly)
Pushrods: Crower ⅜ chrome moly ($140)
Valve Covers: Mopar Performance stamped steel PN P4529339 ($269)
Valve Cover Gaskets: Included with M.P. valve cover kit.
Head Gaskets: Stage V Engineering/Fel Pro ($34 set)
Head Bolt Kit: Stage V Grade 9 ($145)
Spark Plug Tubes: Mopar Performance PN P4120294 ($58)

Camshaft (subtotal = $291)
Type: Mopar Performance solid flat tappet PN P4349344 ($210)
Advertised Duration: 296° intake, 298° exhaust
Lift: 0.572-inch intake, 0.557-inch exhaust
Centerline: 104°
Lifters: Mopar Performance flat tappet solid lifters (supplied w/ cam kit)
Timing Chain: Mopar Performance double-roller, three-bolt PN P5249269 ($81)

Induction (subtotal = $516.45)
Intake Manifold: Mopar Performance M-1 dual plane PN P4452034 ($289)
Intake Gaskets: Stage V Engineering ($12.50 set)
Carburetion: Single Holley 750-cfm, 3310-S ($214.95)

Ignition (subtotal = $301.45)
Distributor: Mopar Performance electronic ignition conversion kit PN P3690428 ($159.95)
Amplifier: Mopar Performance "Orange Box" PN P4120505 (incl. w/kit)
Coil: Mopar Performance performance coil PN P4876732 ($30)
Wires: Mopar Performance 8mm suppression wire set PN P4529032 ($59.50)
Insulators and Boots: Mopar Performance PN P4120808, need 8 ($52)
Timing Specs: 34 deg BTDC
Vacuum Advance: adjustable **HR**

Sources

Dale's Place
Dept. HR10, 5125 Lante St., Baldwin Park, CA 91706; 626/851-1353

Stage V Engineering
Dept. HR10, PO Box 827, Walnut, CA 91788-0827; 909/594-8383; www.stagev.com

Mopar Performance "Magnum *380*" *360*

Test Results

According to Mopar Performance, dyno testing with a 750cfm carb and 1⅞-inch dyno headers show Magnum 380 horsepower output to be 380-plus at 5300 rpm with more than 280 hp from 3800 to 4700 rpm. Torque output is 410-plus lb-ft at 4400 rpm, with more than 350 lb-ft available from 2000 to 4700 rpm.

That's conservative! In our tests, conducted on Dick Landy Industries' SuperFlow dyno, the engine exceeded every claimed output number from 2500 rpm on up (the dyno couldn't hold the engine under 2500). It was the only engine out of the three crate engines tested that surpassed the factory's claimed rating.

As called for, Landy conducted the tests with 1⅞-inch headers and a Holley 750cfm carb, PN 0-3310, which is delivered as a Model 4160 with a fixed secondary metering plate in lieu of replaceable main jets. Landy converted the carb to Model 4150 configuration with replaceable rear jets to facilitate tuning. Final jets were: #65, left-front; #69, right-front; #80 left- and right-rear. Total timing was set at 35 degrees BTDC with Mobil 93-octane unleaded gas using Champion RC12YC plugs gapped at 0.040 inch.

In this configuration, the engine churned out 391.8 hp at 5500 rpm and 418 lb-ft of torque at 3750 rpm! It made more than 300 hp from 4000 rpm up, and more than 300 lb-ft of torque at every rpm tested. Torque output exceeded 400 lb-ft from 3500 through 5000 rpm. What a stump-puller—and a tribute to Chrysler's high-tech Magnum engine designers.

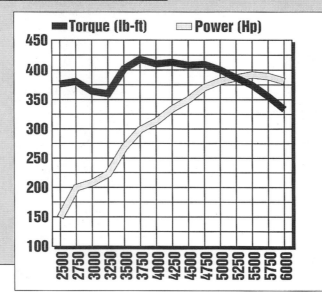

Good Power And Torque Throughout The Rpm Range
By Marlan Davis

Mopar Performance (MP) is replacing its "Super Commando" 360 crate motors with new Magnum-series engines that are based on current-model high-tech pickup truck engines. The top-of-the-line Magnum 380 is conservatively rated at 380-plus hp at 5300 rpm and 410 lb-ft of torque at 4400 rpm. But it's no slouch on the bottom end or midrange, either—horsepower and torque numbers are superior throughout the engine's operating range.

In comparison to the 360hp Super Commando 360 it replaces, the Magnum 380 is up by 20 horsepower. To make 360

A typical factory-style retaining plate that bolts to new lifter valley bosses works with lifter retainers to prevent the hydraulic lifters from rotating within the lifter bores. Although Magnum valvetrains receive oil through the pushrods instead of the block, the old A-engine shaft-mounted rocker cylinder block oil transfer holes *(arrows)* are still present, permitting utilization of traditional shaft-mount rocker-arm heads, if desired.

The new Magnum 62cc double-quench high-swirl combustion chamber *(left)* makes more power with its 1.925-inch intake and 1.625-inch exhaust valves than the old Super Commando did with 2.02/1.60 valves and 72cc chambers *(right)*.

PHOTOS BY MARLAN DAVIS

Mopar Performance "Magnum 380" 360 (PN P5249499)

Displacement:	360 ci (5.9 liters)
Bore x Stroke:	4.0x3.58 inches
Power:	391.8 hp @ 5500 rpm
Torque:	418 lb-ft @ 3250 rpm
Block:	Cast-iron with two-bolt-main bearing caps
Crankshaft:	Production cast-iron
Connecting rods:	Production forged steel
Pistons:	Production Magnum cast aluminum with pressed pins, low-friction $\frac{1}{16}$-$\frac{1}{16}$-$\frac{3}{16}$ rings, 9:1 CR
Camshaft:	Hydraulic roller tappet, 0.501"/0.513" lift, 288°/292° advertised duration, 101.5° lobe separation, 101° intake lobe centerline
Cylinder heads:	Production Magnum high-swirl cast-iron, 62cc chambers, "paired-fulcrum" bolt-down rocker arms with guideplates, 1.925" intake/ 1.625" exhaust valves
Valvesprings:	Heavy-duty single with damper
Induction:	Holley 0-3310 750cfm (converted to Model 4150 configuration)*, M-1 high-rise single-plane aluminum intake manifold
Ignition:	Chrysler electronic distributor, MSD-6 control module*†, MSD Blaster 2 coil*
Exhaust:	1⅝-inch primary tube headers*
Warranty:	90 days/parts from purchase date
Retail price:	$3995 (preliminary, 5/95)

*As tested; not included with engine assembly.
†Engine is delivered with MP "orange box" ECU (PN P4120505).

Magnum engines use a new, shorter cam nose with a correspondingly shorter upper sprocket alignment key *(right)*. Older, long-nose cam billets will clear the Magnum 380's supplied front cover and conventional-rotation water pump, but they will not clear a stock production Magnum engine's short cover and reverse-rotation water pump.

Traditional A-engine intake manifolds won't fit Magnum heads, so MP now offers special intakes just for the Magnum. The single-plane high-rise carries PN P5249501.

Mopar's traditional shaft-mounted rockers are replaced by the new Magnum 1.65:1 paired fulcrum bolt-down rocker arms. Adjustable conversions for the new valvetrain are available from aftermarket sources.

MOPAR PERFORMANCE "MAGNUM 380" 360

horsepower, the Super Commando required higher-compression pistons, larger valves, pocket-ported heads, a high-rise intake manifold and a big hydraulic flat-tappet cam. The Magnum does its thing with just an intake, cam and valvespring change! The heads, block and reciprocating assembly are identical to production '95 360 pickup truck engines.

The new engine is able to generate bigger numbers with fewer mods because Chrysler's basic Magnum engine family is at a higher lever of technological development than past Mopar small-block "A" engines.

Magnum 380 crate motors use high-swirl cast-iron cylinder heads with double-quench combustion chambers and dished pistons that produce a nominal 9:1 compression ratio. Different intake and exhaust port sizing flows more air without mods to the cylinder heads. A high-lift hydraulic roller cam working through the new Magnum paired-fulcrum rocker arm design actuates production Magnum 1.925-inch intake/1.625-inch exhaust valves fed air by a single-plane M1 aluminum intake manifold.

Downstairs, the bottom end remains traditional Mopar 360—two-bolt-main bearing caps, forged rods, and a cast crank and pistons. But in many respects, the internal engine layout still represents what the Bow Tie boys might consider cutting-edge technology.

Example: NASCAR Chevys use a

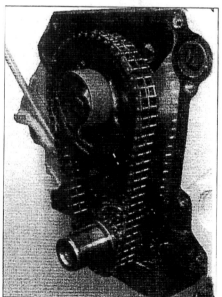

MP's P5249267 high-strength multi-index double-roller timing chain offers decreased rotational friction versus the production silent chain. The sprockets are Magnafluxed for durability. A moly-coated P4120484 eccentric provides mechanical fuel pump compatibility; it's retained by a P4529838 cam bolt and washer set.

Nothing's changed with the Mopar forged con rod—but the 9:1 Magnum '95 pickup truck pistons have a unique rectangular dish that works with the new-design combustion chamber. The pistons utilize low-friction ⅟₁₆-inch top and second compression rings with a low-tension ³⁄₁₆-inch oil ring.

"long" 6-inch center-to-center-length connecting rod in place of the production 5.7-inch length—but stock Mopar 360 rods already measure 6.123 inches. Most race engine builders agree that longer rods make more power with less wear and tear on the cylinder bores.

Example: NASCAR Chevys use 18-degree-valve-angle "rollover" wedge heads for improved flow—but Mopar engines come stock with an 18-degree valve angle that better positions the valves toward the cylinder bore center.

And obviously, stock 360s are 10 inches larger than stock 350s.

Considering that the hottest-ever 360 production combo was '74's 245hp engine, and even the '70 three-two-barrel 340 Six Pack was rated at "only" 290 hp, the new Magnum ranks as the baddest A engine ever released by Chrysler. Now if they'd only build a rear-wheel-drive passenger car to put it in! 🏁

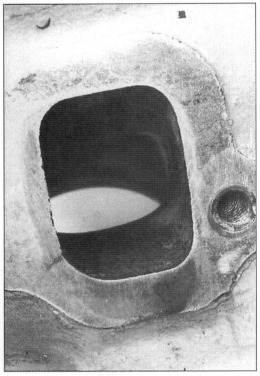

Exhaust port flow is up 30 percent, thanks to the Magnum's improved short-turn radius.

Developing 116 psi of seat pressure at their 1.625-inch installed height, the heavy-duty valvesprings (PN P5249464) fit only Magnum engines, and are used with 2.2-liter four-cylinder engine retainers (PN P4452032) that are compatible with the Magnum's ⁵/₁₆-inch stem-diameter valves (traditional A-engine valves have ⅜-inch stems).

Installation Notes

The Magnum engine assembly does not include a manual trans flywheel or an automatic trans flexplate, a starter, spark plugs (Champion gasketed RC12YC or equivalent recommended), spark plug wires, a fuel pump or an ignition coil. (Part No. P4120889 recommended if using the "Orange Box" ECU shipped with the engine.)

Like earlier 360 engines, the Magnum 380 is externally balanced. However, the amount of external unbalance differs: '71-'92 360s use 19.79 in-oz balance, but all Magnum 360s (both stock production and the high-perf crate engines) use 14.65 in-oz of external balance. This means that existing "traditional" 360 production and MP manual trans flywheels, production 360 auto trans torque converters, or MP's current high-perf converter weight and balance package won't work on Magnum engines without rebalancing. (Flexplates interchange, because Mopars have the unbalance weights on the converter.) MP is developing manual trans flywheels and auto trans torque converter weight packages for the Magnum 360, and they may in fact be available by the time you read this.

The Magnum block includes early-style mount pads, allowing the engine to bolt in place of earlier 340 and 360 engines. However, 273/318 engines use a different driver-side engine mount, which must be changed to a 340/360-type mount. Big-block and slant-six-series engines have totally different mounts and bellhousing bolt patterns, so converting from them to a 360 requires changing the trans, the mounts and (in some cases) the K-member.

Magnum 380 engines come with a mechanical fuel-pump-compatible front cover and standard-rotation V-belt-compatible water pump. The current stock production Magnum front

The late Magnum block includes both early- and late-style engine mount lugs. The 273/318 driver-side engine mount is different from the 340/360 mount.

cover (PN 53006704) and serpentine-beltdrive reverse-rotation water pump (PN 53020280) is at least 2¼ inches shorter (not counting the more compact serpentine drive itself), which would provide additional clearance on custom swaps, but you'd need to run late-model brackets, accessories and an electric fuel pump.

The Magnum cylinder heads represent a complete departure from earlier A-engine cylinder head designs. Old exhaust manifolds and headers will fit. All accessory holes are drilled in the ends of the heads. Non-Magnum intakes *won't* fit. The Magnum valvetrain is unique. Swapping typical shaft-mounted rocker A-engine heads requires changing everything from the lifters on up. Likewise, non-Magnum Chrysler lifters won't work with the Magnum heads and valvetrain, because the new valvetrain oils through the pushrods, but the shaft-mount system oils through the block. The block still has the old oil transfer holes, however. You can install conventional A-engine flat-tappet hydraulic, solid lifter or solid roller lifter cams with Magnum heads by using aftermarket American Motors V8-type lifters (convert to adjustable valvetrain, as required).

All production Magnum engines are fuel-injected, so the heads have no exhaust heat riser provisions. Carbureted applications should use electric chokes with a remote exhaust-mounted heat sensor. While this engine exhibits a broad torque and power curve, the extremely tight lobe separation may cause a vacuum deficit, requiring a vacuum reserve canister if you have power brakes. A 3000-stall torque converter is highly recommended. This assembly is sold for off-road use only.

The front cover accepts mechanical fuel pumps and is designed for a standard-rotation, V-belt-style water pump

MP's 360 oil pans are different and won't interchange with 273/318/340 pans. The as-

sembly's center-sump oil pan *(shown)* fits most passenger-car applications. Two- and four-wheel-drive pickups and Ramchargers require a rear-sump pan (PN P5249060; includes pan, pickup, dipstick and tube).

Sources

(Test facility)
Dick Landy Industries
Dept. CC
19743 Bahama St.
Northridge, CA 91324
818/341-4143

(Product source)
Mopar Performance Headquarters
Dept. CC
P.O. Box 360445
Strongsville, OH 44136-9919
810/853-7290, tech line

MAXIMUM MOPAR

POWER FAST & EASY

By John Baechtel

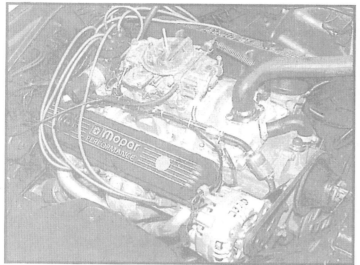

Thanks to the strong factory development program at Mopar Performance, 318 to 360 small-blocks offer some of the best potential power packages around.

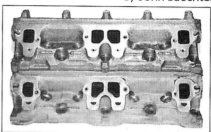

The new high-swirl 318 head (bottom) features raised exhaust ports with 1.88-inch intake valves and 1.60-inch exhaust valves. High-swirl chambers produce considerable power on all small-block applications.

The trick to building a good, strong Mopar street engine is selecting the proper mix of Mopar Performance and aftermarket parts that will make an engine powerful *and* affordable. Fortunately, Chrysler's A-series engines are blessed with a factory Mopar Performance development program that provides up-to-date performance hardware.

Like the other engines in this section, the Chrysler A engine responds well to modifications that enhance breathing. Using the ideal street compression ratios of 9.0:1 to 9.5:1, you can produce 325 hp just by pushing the engine out of a tree. It takes some pretty serious modifications to crank 400-plus hp out of a streetable 318, but a 400-hp 360 can be assembled with off-the-shelf parts. Any well-built short-block can deliver these power levels if it's equipped with the proper breathing components. Forged pistons are always desirable because they're more durable, especially if you plumb the engine with nitrous oxide. Cast cranks and two-bolt blocks are also acceptable, as long as you perform accurate machine work and ensure proper lubrication.

You can even build 300- to 400-hp engines by using old-technology performance parts, which may be the most economical approach. Earlier cylinder heads are capable of delivering the desired power levels, so don't overlook them if you find a good deal. Chrysler's 318 ported head (part No. P445-2748) is an example of current technology. It is based on the '87 swirl port design and features 1.88-inch-diameter intake valves and 1.60-inch exhaust valves. If you're working with a 340-cubic-inch engine, the new 340 T/A and AAR head, which is based on a 360 casting, is a worthwhile investment. These heads have hardened exhaust seats, 2.02-inch intake valves, and pushrod holes to match the original 340 T/A and AAR heads.

There are two inexpensive cylinder heads available for the 360-cubic-inch A engine. Both are high-swirl, fully cc'd, ported heads with 2.02- and 1.60-inch valves. They come completely assembled with performance springs and chrome-moly retainers and are compatible with standard intake and exhaust manifolds. Part No. P4529588 is suggested for cams with up to a .500-inch lift, and part No. P4529589 is recommended for cams with over a .500-inch lift.

Selecting a camshaft for Mopar engines is easy because Mopar Performance has removed all the guesswork. It offers specific recommendations for all performance levels from 17 seconds down to the mid 10s. There are also specific recommendations for manual and automatic transmissions, and all compatible valve gears are listed in the Mopar Performance catalog that can be purchased from any local dealer. If you want to select an aftermarket cam, follow the guidelines suggested in the accompanying chart. However, for most applications, you can't go wrong with Mopar Performance recommendations. All Mopar Performance cams are designed and track tested by Chrysler engineers, and each camshaft kit comes with a certificate for a free VHS video tape that details a step-by-step cam installation and centerlining techniques.

Continued on page 121

318-360 CHRYSLER		
	300 HP	**400 HP**
CARBURETOR	600 to 750 cfm Holley 4V, Rochester Q-jet, Carter AFB	700 to 750 cfm Holley 4V, Rochester Q-jet, Carter AFB
INTAKE MANIFOLD	Edelbrock Performer 2176, Weiand 925-8007, Mopar Performance W-5 P4529295 or P4529116 (non W-5)	Edelbrock Torker II 5076, Weiand 925-7545, Mopar Performance W-5 P4529295 or P4529294
CAMSHAFT (Hyd) Duration (@ .050) Intake Exhaust Lift Intake (in.) Lift Exhaust (in.)	204° 204°/216° .420/.427 in .420/.454 in	222°/224° 232° .447/.467 in .450/.494 in
HEADERS Tube dia. (in.) Collector dia. (in.)	1⅝ in 3 in	1¾ in 3½ in
HEADS Intake Valve dia. Ex. Valve dia. Part Nos.	1.88-2.02 in 1.60-in P4529268 (318) P4529269 (360) P4529493 (340) P4529588 (ported 360 high swirl)	2.02 in 1.60 in P4452748 (318 ported) P4529589 (ported 360 high swirl) P4452924 (W-5 aluminum)

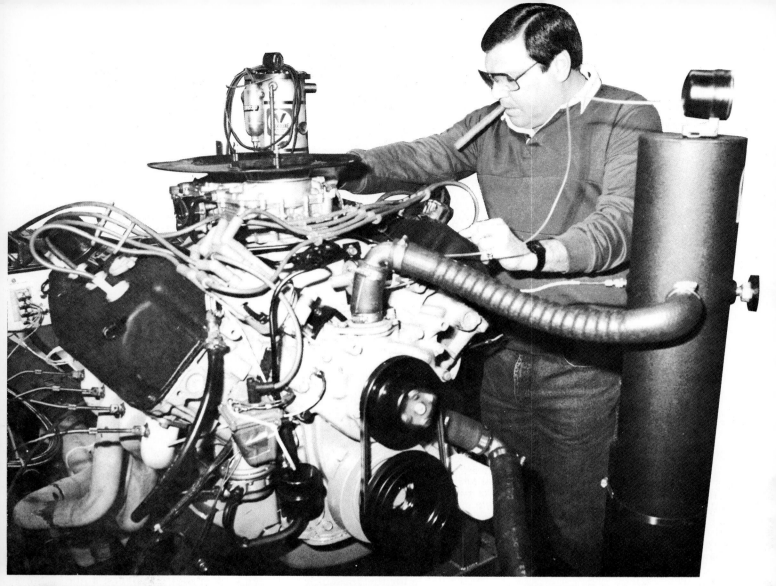

STREET HEMI

By Kevin Boales

Chrysler's Hemi engines are magic. In addition to incredible visual effect under the hood, their real charm lies in the performance potential offered by the chamber shape—a shape which has been in existence since before Weslake's time and one that remains today inside the world's fastest Top Fuelers, Funny Cars, and some Pro Stocks. From its introduction in the early Fifties, the Chrysler Hemi has been a controversial piece of work. Like most engineering triumphs, however, the Hemi has two peculiarities: it burns a lot of oil and it makes a lot of valvetrain noise. Like the bionic man, today we can approach these problems with the attitude: "We can fix it; we *have* the technology!"

THE FINAL FRONTIER: A STREET HEMI THAT DOESN'T BURN OIL OR RATTLE

Actually, Dick Landy, long the Pro Stock Hemi guru to dozens of winning efforts through the years, has the technology. We dropped by his shop to see just how he solves the oiling and rattling for the street. He complied by taking us through Richard Artunian's gennie hemi, to be installed in a completely restored Superbird (which we can't wait to climb all over when it's ready to photograph).

The oiling problem stems from the use of forged pistons in the Hemis built in the late Sixties. Forged pistons are made from a different alloy than cast ones, and the forgings expand more than the cast pistons. Because of their expansion rate, forged pistons are installed with more clearance in the bores, leading to the noise *and* oiling problems while the engine is warming up. Dick's solution to this dilemma is similar to the approach the Duesenberg brothers used 50 years ago; he thermally isolates the top of the forged pistons in his street engines by cutting a slot through the oil control ring groove extending nearly 90 degrees around the piston on both sides parallel to the wrist pin. The Duesenberg fix was to cut a groove around the entire circumference of their cast pistons, leaving only the pin bosses to tie the top and skirt of the pistons together. The result is that the pistons can now be fitted to the bores

HEMI

with "cast" piston clearances. Landy uses .003-inch, and the Duesenbergs' castings were fitted at .0015, again leaning on the lower expansion rate of the castings.

DLI (Dick Landy Industries) stocks most of the parts you'd need to build a street hemi, including the DLI/J&E Forgings used in Artunian's engine. The pistons are supplied with a "J" taper, which denotes a skirt taper of .004-inch from top to bottom. Since more heat is available to expand the skirt at the top, it's cut smaller than the bottom. The slugs are also cam ground to compensate for added growth at the pin bosses; that portion of the skirt measures .021-inch smaller than a measurement taken perpendicular to the

The only head modification visible here is a very slight cleanup around the intake and exhaust seats to help top-end flow. Chamber shape allows stout 10.75:1 ratio to work on pump gas.

For the street, there's not too much wrong with stock valves, retainers, and locks. K-Line guide bushings and DLI seals fix inherent oiling problems.

Stock cast-iron exhaust manifolds are adequate for the street, and better than some tubing versions at low rpm. Since this is a restoration, they're the only choice.

J & E/DLI forgings are slotted to thermally insulate head from skirt, allowing the piston/wall clearance to shrink from .008-inch to .003-inch, reducing noise and improving ring seal.

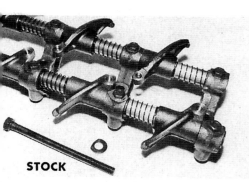

STOCK

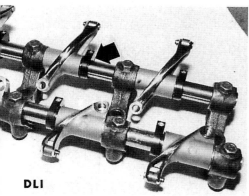

DLI

DLI rocker arm setup is a product of Top Fuel evolution. It eliminates a serious valvetrain problem caused by lateral motion of the stock system.

Specification Table

Item	Size	Clearance	Mfgr.
Crankshaft	3.750-inch stroke		Chrysler
Rod journals	2.374	.0025	
Main journal	2.749	.003	
Rod side clearance		.012	
Thrust brg. clearance		.004	
Rod big end	2.500 (machined size)		
Rod small end	1.032 (finished size)		
Rod length	6.861		Chrysler
Wrist pin	3.400x1.031 x.250 wall		Chrysler
Block deck height	10.72 inches		
Block main bore diameter	2.9425		
Main caps			Chrysler
Cylinder bore	4.2555		
Piston diameter	4.2525 (largest measurement)		DLI/J&E #1289
Piston/wall clearance		.003	
Piston wrist pin diameter	1.031	.0007	
Piston cam	.021		
Piston taper	.004		
Rings	5/64, 5/64, & 3/16 Speed Pro #R9590		
Wrist pin retention			dual Truarc
Pin end play		.005	

Camshaft duration: 242-degree intake; 252-degree exhaust
Cam centerline: 112 degrees
Camshaft lift: .553 intake; .559 exhaust
Cam grind number: Crane H-242/3520-2-12
Valvespring: Chrysler Hemi p/n 3690933
Seat pressure: 115 lbs. @ 1.870
Open pressure: 360 lbs. @ .10" from coil bind
Valve keepers and retainers: Chrysler

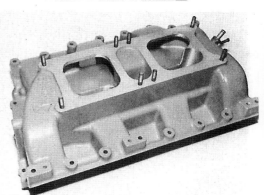

Stock dual-quad manifold will support two Carter AFB's. Divider is machined from beneath carb flanges to enhance airflow characteristics in the top end.

Mike Landy uses two torque wrenches to handle the chores; beam-style wrench is used to finish all the bolts (and nuts in intake valley).

Crane cam's dual-pattern Chevy big-block pattern, H-242/3520-2-12, was chosen and ground onto Hemi billet. This pattern works well with the Hemi valvetrain.

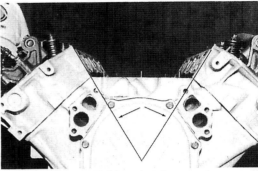

Steep intake manifold gasket face angle is primarily necessitated by valve inclination. Overtorquing intake bolts will damage the heads in log splitter style.

DLI modifies the oiling system to include this low restriction pickup, a windage tray, flow limiters in the rocker stand galleys, and a high-flow pump.

Fel-Pro's Pro Lock compound is used to secure the splash pan below the intake manifold. This is cheap, effective insurance against small hardware wandering around inside the engine.

The windage tray needed a little help to clear the crank cheeks, so Mike adjusted it with the proper tool. Gasket flange portion is clamped to bench.

HORSEPOWER DLI ROCKERS   STOCK ROCKERS

500
400
300

3500 3750 4000 4250 4500 4750 5000 5250 5500 5750 6000 6250 6500
ENGINE RPM

TORQUE DLI ROCKERS   STOCK ROCKERS

440
420
400
380

3500 3750 4000 4250 4500 4750 5000 5250 5500 5750 6000 6250 6500
ENGINE RPM

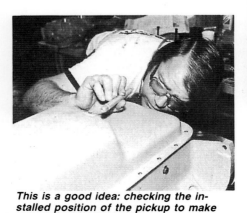

This is a good idea: checking the installed position of the pickup to make sure it's far enough away from the pan floor to allow free oil flow.

You can see from this graph the difference in performance achieved by stabilizing the valvetrain on the Hemi. You can also see why the NHRA factors the Hemi in at 480 horses for the Stock Eliminator class competitors. Stock Hemi engines, without oiling problems, will make the advertised 425 with a plug wire pulled. . . .

LANDY HEMI

pin. DLI also fly-cuts and lightens the forgings for each particular application. Reducing the piston-to-wall clearance allows the rings to remain in contact squarely with the cylinder wall during warm-up instead of losing the seal as the piston rocks around; obviously, piston scuffing problems are also minimized with tighter tolerances.

There are other sources of noise lurking inside each Hemi. DLI has found that the valvetrain on these engines, an incredible and unlikely assortment of weird rockers, springs, and shafts, not only makes enough noise to wake up small Midwestern towns, but can actually cost you horsepower by changing the closing pattern of the camshaft through lateral motion of the rockers on the shaft. The way the rockers are loaded against the valves and springs causes them to be pressed against the stands while the valve opens, but as it closes, the screwball geometry problem created with the design causes the rockers to move away from the stands, effectively altering the closing rate of the valve. The sure cure for this situation was no further away than DLI's Top Fuel rocker system, modified slightly to make pushrod installation easier. With the DLI rocker setup, the rockers can't move away from the stands because they're retained on the shaft by a locking collar. Quieter operation and a more predictable closing action are guaranteed; the horsepower you lost with the stock rocker setup can be seen in the dyno chart. Keeping the rocker under lateral control also eliminates the possibility of scuffing or side loading the valve stem, a situation that in the past led to premature guide failures and even broken valves.

With this newfound accuracy in the valvetrain, the guides can also be tightened up around the valvestems. DLI uses a special K-Line bushing and sets up the clearance to the stem at zero. None. Squeak. The valves literally clearance themselves when the engine starts for the first time. The stem seals are replaced with a DLI-custom conical Teflon seal that allows oil flow along the stem until pressure inside the engine block and valve covers gets to a certain point, at which time the oil flow is reduced to almost zip. Perfect Circle seals, used on most other high-performance engines, are a little *too* effective at oil control for use with the no-clearance guide bushings on DLI's engine builds. Using the K-Line bushings and DLI's seals limits the amount of oil traveling down the valvestems just enough to lube the valve, but not enough so that any oil will creep into the chamber after shutdown or during warm-up.

Last on the list of oil control items is a high-energy ignition system, to keep the fire lit in case the oil manages to get by the other tricks in the engine. DLI usually recommends Direct Connection's electronic ignition modules and distributors for street Hemis; the latest vacuum advance distributor and control module is their PN P3690428 for the 413/426/440 wedge and Hemi engines, with a mechanical advance distributor and race-only control module identified by PN P4286512 also available. In either case, cold-start problems are a thing of the past. On 1969 or earlier cars, when you install either of these systems, you'll need to replace the voltage regulator with PN P3690732 to handle the new electrical loads properly, and if you use the mechanical system, you'll also need an Accel coil, DC PN P3690560.

Most of the top end stuff is stock Chrysler, with the noted exceptions. The Crane cam grind for this engine is actually a Chevy BB split pattern that happens to suit the Hemi perfectly. Transferred to a Hemi billet, the pattern establishes the right sequence and rate for the Hemi valvetrain with no valve/piston interference problems. Juice lifters are used for simple maintenance and additional noise control.

The remainder of this street Hemi is necessarily stock, since Artunian is restoring the car from nose to wing. The dual-quad Hemi manifold is modified internally by removing part of the plenum divider, effectively making the manifold resemble a single-plane design on the dyno. By knocking out the divider, each port has access to all eight venturis rather than only four. Dick claims this "fools" the engine into thinking it has more carburetion, resulting in a top end increase through improved breathing. Leaving enough of the dual-plane webs in place is important for low-end performance, despite the fact that the Hemi doesn't really need too much help in the torque department.

Due to the acute angle formed by the intake manifold mounting faces on the cylinder heads, it's important to keep an inch-pound torque wrench handy when you install the manifold. Like a log splitter, *it is possible* to pull chunks of the heads out by overtightening the intake bolts. That's one of the reasons the Chrysler engineers used such small hardware to attach the manifold; they'd rather you broke a bolt than scrapped a head.

The tables show the performance of the finished engine with and without DLI rockers. What the tables don't illustrate is how much quieter the engine runs with them. If you couldn't see those great big valve covers with the plug wires heading straight down the middle, you'd never guess it was a Hemi...until you hit the throttle. **HR**

Continued from page 117

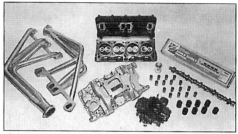

A low-buck combination that uses stock 360 heads with headers, an Edelbrock Performer intake, a 600-cfm vacuum secondary Holley, and a factory camshaft will yield 300-hp from a 360.

The same can be said when selecting an intake manifold. Mopar Performance designers have taken the guesswork out of intake selection by offering a full range of dual-plane and single-plane intakes. The standard 4-barrel dual plane (part No. P4529116) is the ideal starting point for all A-engine buildups. It easily delivers 325-hp on all size engines and can approach the higher power levels on larger 360-cubic-inch engines. It is most effective with applications that do not run W-2 or W-5 cylinder heads. Selecting a carburetor for these engines is similar to standard-cfm-versus-engine-size matching procedures (see chart). Smaller engines will respond better with 600- to 650-cfm carbs while larger engines can use a 750-cfm carb. With the proper combination of parts, you can even get close to 400 hp from a 318-based engine, using no more than a 650-cfm double-pumper with a factory cam, heads, and manifold.

All of these combinations will respond to appropriate header selection and a free-flowing exhaust. Milder engines should use 1⅝-inch primary tubes with 3-inch collectors, but larger displacement engines and extra-strong 318s will benefit from 1¾-inch primaries with 3½-inch collectors. Don't overlook Mopar's high-energy ignition systems, and give all these engines as much breathing capability as possible by incorporating a high-flow air filter. Like the Ford and Chevy small-block, the Chrysler small-block has been delivering strong, reliable power for a long time. The latest offerings from Mopar Performance represent the most current technology, and they will yield the best results when building a Mopar street engine. **HR**

SOURCE

Mopar Performance
Dept. HR
P.O. Box 215020
Auburn Hills, MI
48321-5020
(313) 853-7290

Armed And Dangerous

Build a 560hp, 478ci Chrysler Hemi Stroker!

By Marlan Davis

Installing a crank with a longer stroke than the original production crank is a tried and true method for gaining a lot more cubic inches than is possible with a simple overbore. It's also a favorite technique of take-no-prisoners Pro Streeters. But Jack Falkenrath's Chrysler Hemi Elephant motor was originally a straight restoration project for his '70 Challenger; he would have been satisfied if the car ran 12-second quarter-mile times, an e.t. any properly tuned stock Hemi with slicks is capable of accomplishing. So why would Falkenrath consider going the stroker route?

The answer is obvious to anyone who has ever driven a street Hemi. The race-inspired design's large ports require high rpm to make the engine run efficiently. With a longer-stroke crank, the air/fuel column within the port achieves higher velocity at low rpm, which in turn generates more torque. And as anyone who has driven a 5.0L EFI Mustang or newer LT1 Corvette can tell you, tire-twisting torque makes a car a helluva lot more fun to drive!

With the goal of becoming the torque of the town now firmly in mind, Falkenrath had the engine delivered to the ace Hemi builders at Dick Landy Industries (DLI), who handled the complete engine-machining process. The original block cleaned up with a 0.031-inch overbore. Combine that with the Mopar Performance stroker crank's 4.15-inch stroke (compared to 3.75-inch for a stock Hemi) and you get 477.9 ci.

After cleaning and measuring all critical components, the DLI team ordered custom JE reverse-dome pistons.

The stroker crank dictated moving the piston pins 0.200 inch higher in the piston compared to the stock Hemi pin location. The pistons also have extra material around their outside dome edges, which allowed DLI to custom-machine and then hand-fit each piston dome to within 0.060 inch of its corresponding cylinder-head combustion chamber. The resulting near-collision of the piston's outside edge with the chamber further atomizes the air/fuel mixture, making it possible to run slightly higher compression ratios without encouraging detonation. While close-fitting reverse-dome pistons are commonly used on race engines, this is the first time DLI used this type of piston in a street Hemi engine.

Calculating Hemi compression ratios is tricky. A Hemi piston has no piston deck (flat) area, so fluid dispensed from a chemical burette is used to calculate the actual dome displacement with the piston positioned 1 inch below TDC. The amount of fluid required to fill the cylinder to the top of the block is subtracted from the area of a 1-inch-long cylinder that has a diameter equal to the engine's bore. Because it is incorporated into the TDC piston dome volume product, no separate piston deck height volume calculation is needed. Instead, DLI mills the block to achieve a desired dome top-to-block deck height. With an actual 105.9cc TDC volume (178cc cylinder-head combustion chamber + 76.9cc piston dome volume – 4.8cc gasket volume), the final com-

Premium bottom-end components include Mopar Performance's stroker crank (balanced with Mallory heavy metal), main bearings, and rod bearings. Speed-Pro supplied the file-fit piston rings, while JE forged the custom pistons. Like the block, the beefy production Hemi rods are a carryover from the previous engine—but they were thoroughly Magnafluxed, shot-peened, polished, sized, and fitted with ARP bolts before reuse.

The reverse-dome JE Hemi piston (*right*) is worth about 20 lb-ft of torque over the conventional stock-type Hemi piston (*left*). Common on today's race Hemis, the reverse-dome design generates "squench" (a combination of squish and quench) in the chamber. This generates increased turbulence at TDC.

When utilizing quench-head pistons in a Hemi, it is imperative that the combustion chambers be perfectly centered over the cylinder bores. If they're not, the pistons will contact the head—with disastrous consequences. Here, DLI visually checks for the proper alignment. If necessary, the head dowel pin locations are corrected to blueprint locations by offset-machining the block and heads for oversize pins.

To promote cylinder sealing, this engine uses 0.005-inch oversize Speed-Pro rings. This allows DLI to custom-file the end-gaps to the optimum size for each bore (0.016-inch top, 0.012-inch second). DLI honed the block using torque plates to simulate the distorting effects of a torqued cylinder head, which ensures the rings actually run in straight and round cylinders.

Mods to the original Hemi intake manifold include cutting down the plenum divider and radiusing the runner entries (arrow). These changes for increased top-end horsepower are an updated version of Chrysler's '60s racer recommendations.

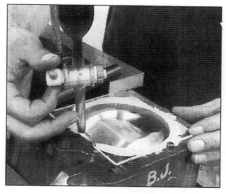

On a Hemi, builders check effective piston dome volume with the piston 1 inch below TDC. A Plexiglas plate is sealed to the top of the cylinder bore with grease, and a burette is used to fill the cylinder bore to the top of the block with colored fluid.

DLI degreed in the cam using the intake centerline method at 107 degrees ATDC. Hi-Tech's Acu-Tech Hemi True Roller timing set has an endlessly riveted, solid-bushing, BSI-style chain; a larger tooth-profile, high-tensile cast-iron, three-bolt cam sprocket; and a three-keyway SA-1144 stress-proof steel crank sprocket that's hardened to Rockwell C-50 minimum.

pression ratio is 10.25:1...and yes, it runs great on pump gas!

A custom Hi-Tech Engine Components hydraulic cam actuates the valves installed in the original Hemi heads, which received mild headwork before being bolted in place atop Mopar Performance steel-shim head gaskets. The valvetrain's durability is improved by replacing the original Hemi rocker arm retainers with DLI billet aluminum retaining collars. Chrysler valve-stem seals originally designed for '93-and-up Magnum 360 truck engines are used on this Hemi; their Fluroelastomer (Viton FKM) material is more durable than the OEM Teflon seals. Hi-Tech offers this new-material seal for this and many other applications besides the Hemi.

Once assembly was complete, the engine was off to Landy's engine dyno. After final tuning, the twin Carter AFB carburetor-equipped engine responded with 560 hp at 5,500 rpm and a whopping 570 lb-ft of torque at 4,500 rpm. Not only did the special Hi-Tech cam grind out incredible peak numbers, but it also generated a super-flat curve as

well: More than 500 hp was produced between 4,750 and 6,000 rpm (the highest rpm attained during testing), and the engine made over 500 lb-ft from 3,000 rpm on up. Combine these numbers with 4.10:1 rear gears and a four-speed, and you've got one traction-challenged Challenger. This ain't no elephant walk...it's a stampede!

Mopar Performance chrome-vanadium valvesprings and chrome-moly 8-degree retainers return huge 2.25-inch intake/1.94-inch exhaust valves to their seats. Stock cast rocker arms do the job (Hemi roller rocker sets cost over $1,000). DLI billet aluminum rocker arm spacer plates, '93-and-later Magnum 360 truck Viton valve seals, and valve-stem lash caps increase durability.

DLI machines the heads for the Hi-Tech manganese-bronze valve guides custom-made for this engine. These guides offer superior thermal conductivity, a low friction coefficient, excellent corrosion resistance, and compatibility with no-lead, propane, and alcohol fuels.

The Elephant Man

Dick Landy was a phenomenally successful A/FX, Super Stock, Pro Stock, and match-race racer during the '60s and '70s. Because he was so instrumental in the 426's racing development, Landy's name has become synonymous with the Hemi engine. In fact, Landy even worked for Chrysler during the musclecar era, conducting performance clinics around the country, and he is still involved in the development of parts for Chrysler's rereleased Hemi engines. Landy has never lost his hands-on attitude.

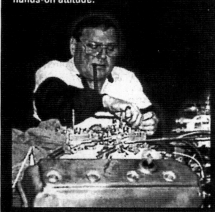

Hi-Tech Engine Components' special hydraulic cam grind features 0.528-inch intake/0.555-inch exhaust valve lift; 232-degree intake/242-degree exhaust, 0.050-inch tappet lift duration; and a wide, 112-degree lobe separation angle for less reversion with the stroker crank and a broader torque curve.

DLI made the awesome Hemi cylinder heads even better by blending the bowls and cleaning up the combustion chambers.

And in This Corner, the Challenger...

While Landy was resurrecting the Hemi, other items on Falkenrath's rare '70 Hemi Challenger were being restified. Upgrades include a late-model quick-ratio power-steering box, Mopar Performance gas shocks, modern Goodyear Eagle ST radial tires, urethane sway-bar and control-arm bushings, Mopar Performance electronic ignition, and a Lakewood steel scattershield for safety.

Because the car will be shown at car show Stock classes, it was important to keep it looking as original as possible. Luckily most of the changes are hard to see. For example, the new power-steering box looks just like the original, and because the gas shocks have the same external dimensions as the OEM units, a quick coat of black paint made them look just like the factory stockers. In fact, the restoration even went to the extent of reproducing all the original factory paint codes onto the new or repainted front-end components.

The Challenger is one of only 112 made with a Hemi and a four-speed in 1970. With just 45,000 miles showing on the odometer, the car's original paint still looks sharp.

The reproduction battery makes the car look just like it did the day it left Chrysler's Hamtramck (Detroit) assembly plant. Year One provided hard-to-find original-style parts—including the battery, 15-inch wheel trim-rings, fuel lines, motor mounts, wiring harness, exhaust tips, and Hemi-only steering pulley. The rare power-steering bracket and cooler were located in a local salvage yard.

Engine builder and dyno operator Mike Landy prepares to tweak the 478 Hemi. After some intake manifold mods and carb tuning, the engine made 570 lb-ft and 560 hp. **CC**

426 HEMI DYNO TEST

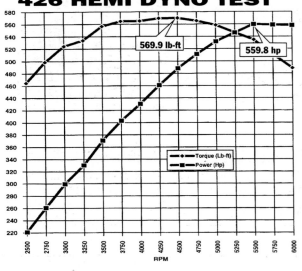

569.9 lb-ft

559.8 hp

Legend: Torque (Lb-ft) / Power (Hp)

RPM

Sources

Automotive Racing Products (ARP)
Dept. CC
531 Spectrum Cir.
Oxnard, CA 93030
805/278-RACE

Dick Landy Industries (DLI)
Dept. CC
19743 Bahama St.
Northridge, CA 91324
818/341-4143

JE Pistons
Dept. CC
15312 Connector Ln.
Huntington Beach., CA 92649
714/898-9763

No. 1 Performance
(*Mopar Performance & Hi-Tech Engine Components distributor*)
Dept. CC
1775 S. Redwood Rd.
Salt Lake City, UT 84104
800/453-8250

Year One Inc.
Dept. CC
P.O. Box 129
Tucker, GA 30085
800/950-9503